Translator's note

When I was first asked to translate the German edition of this book, I was aware of some of the work of Prof. *Stegemann* and his German colleagues. As I read through the book, however, I realized that there was a great deal of German research with which I was not familiar and that the majority of the information contained therein was not readily available in the English scientific literature. For that reason alone, I felt that the book was worthwhile to translate. In addition to the book's informational content, I found that some of its concepts and ways of thinking about exercise physiology were different from those normally found in English books on the same topics. In particular, I found Prof. Stegemann's use of cybernetics and its concept of man as a machine both interesting and unique.

Since communication is important in the progress of science, I hope that this English edition will help expand the knowledge of its English-speaking readers and that they will better appreciate the research being done by their German-speaking colleagues. If this happens, then this translation will have performed a service.

London, Ontario, Spring 1981   *James S. Skinner*

Contents

Exercise Physiology

Physiologic Bases of Work and Sport

Jürgen Stegemann

Translated by James S. Skinner

197 Illustrations, 25 Tables

1981
Georg Thieme Verlag Stuttgart · New York

Author

Professor Dr. med. Jürgen Stegemann
Physiologisches Institut der
Deutschen Sporthochschule Köln
Carl-Diem-Weg
5000 Köln 41, FRG

Translator

James S. Skinner, Ph. D.
Professor
Faculty of Physical Education
The University of Western Ontario
London, Canada N6A 3K7

Deutsche Bibliothek Cataloguing in Publication Data

Stegemann, Jürgen:
Exercise physiology : physiologic bases of
work and sport / by Jürgen Stegemann. Transl.
and ed. by James S. Skinner. – Chicago ; London :
Year Book Medical Publ. ; Stuttgart : Thieme,
1981.
  Einheitssacht.: Leistungsphysiologie 〈engl.〉

1st German edition 1971
2nd German edition 1977

© 1981 Georg Thieme Verlag, Herdweg 63, D 7000 Stuttgart 1, FRG

Typesetting by Dörr, Ludwigsburg on Linotron 202
Printed in West Germany by Dörr, Ludwigsburg

ISBN 3 13 582501 9   5 4 3 2 1 0

...achers

Foreword to the Second German Edition

Exercise physiology is continually changing, not only because of new research findings but also because of its main areas of interest. Sports physiology is assuming greater importance, while classic work physiology is assuming less. This is primarily because the major problems in the working place of today are psychologic rather than physiologic. At the same time, health consciousness has increased, along with the associated rise in athletic activities.

This second edition has tried to take this development into account, especially in light of the fact that the pocket book is increasingly being used as a textbook for the physiologic instruction of physical education students. These readers wanted to have an amplification of the sections on fundamentals and the inclusion of a supplement on terminology. In addition to these requests, a new section on motor activity and motor learning has been included. The description of blood gas transport and of the circulatory system, as well as the regulation of ventilation during exercise, has been expanded. A special section has been devoted to the problems of weightlessness and their relationship to performance capacity, orthostatic tolerance and other factors. The physiologic bases of different forms of training have been completely reworked.

To clarify and assist the understanding of the cybernetically described relationships, the basic principles of biological regulation are discussed in the supplement. Another supplement details methodologic problems.

I am very grateful to Prof. Dr. *F. Hammersen* for assistance with the description of muscle structure and I am indebted to him for the use of the accompanying electron microscopic photographs. I am also grateful to my associates for many years, Prof. Dr. *D. Böning,* Dr. *U. Tibes* and Dr. *W. Skipka* for their critical review of the manuscript and their many suggestions for improving the presentation. I would like to thank Dr. *W. Skipka*, Mr. *H. Boden* and Mrs. *J. Jagonak* for their support during the compilation of the terminology and index of key words and Mrs. *Petra Dzwonnek* for her untiring technical assistance. I am especially obliged to the many colleagues and students who have performed a real service by their helpful criticisms and stimulation.

Finally, I would like to thank Dr. h. c. *G. Hauff,* and the Georg Thieme Verlag for their constant, friendly cooperation.

Cologne, Spring 1976   *Jürgen Stegemann*

X   Contents

Introduction

Exercise physiology encompasses two large areas of applied physiology: physiology of occupational work (also called work physiology) and sport physiology (principally the physiology of high-level performance). Both areas have common and different elements which will be discussed in detail throughout this book.

Up to the present time, physiology, which is the natural science of the function of living, has been studied and considered in its totality in only a few countries. Today, various aspects of physiology are studied by numerous scientists from totally different disciplines with common interests overlapping in the field of physiology. The main disciplines are: medicine, which applies knowledge of physiology to the problem of pathologic states and attempts to improve its understanding in this way; zoology, which generally is interested in the comparative aspects of physiologic functions; and for about the last 50 years, the field of physical work and sport, which applies the knowledge of physiology and its methods. In the case of work physiology, the primary interests are the protection of the worker from injuries due to overwork, as well as logical physiologic explanations and the influence of environmental conditions on the worker.

The world of sport is interested in physiologic knowledge because physical stress is particularly important in sport. In competitive sport, the athlete strives to use his knowledge of physiology to improve his performance. As long as this knowledge serves as the basis for training methods that are founded on physiologic principles and as long as athletes train according to this knowledge, it is clearly a positive development. If they use this knowledge to stress the body beyond its physiologic limits of performance, e. g., by means of pharmaceutical agents (doping), then the development is hazardous. To the physiologist, high-performance sport is a source of knowledge about the foundations of physiology, because extremely high-level athletic performance is also an experiment. For example, it is not difficult to imagine the major stimulus for research that has come from the Olympic games. Physiologists have much to offer and to receive from the physiology of sport.

Work Physiology

Work physiology is part of the physiologic sciences and as such, its comprehension presupposes a basic knowledge of human physiology.

Work physiology can be considered a branch of physiology with a special orientation. It is necessary to view the physiologic consequences from a definite viewpoint, i. e., the conditions of occupational work. As a result of that description, the major points of emphasis among the individual areas in physiology must be shifted to the organs and organ systems which will be stressed under conditions of physical work or which will limit the work that can be done. Systems whose performance is barely activated during exercise are ignored using this way of thinking. Therefore, work physiology is pursued by physiologists whose interest is associated with this special way of viewing physiology. For this reason, it would be incorrect to characterize work physiology primarily as being exclusively applied research, whose task is purely practical. Work physiology is much more like many other branches of physiology, in that it is also a basic science. Physical work done by man is also the object of natural scientific research for physiologists.

A. Problems and Tasks of Work Physiology

Science and research do not exist in space without reference to other things. Work physiology has an especially important social task, i. e., to humanize man's working place as a branch of ergonomics. Work physiology is simultaneously fundamental and applied research. It is basic research because its responsibility is to systematically investigate the interplay of the organs and organ systems; it is applied research because it applies the knowledge about this interplay to adapt the world of work to the functions of man, i. e., to rationalize it.

In the sense of work physiology, the term "work" is defined in physics as the product of force and distance. However, this definition describes only a small part of what is generally called work in colloquial speech. Colloquial speech does not differentiate between work as the result of an activity and the activity itself. Work, in the sense of work physiology, is a relatively broad term. It is primarily the activity which man exerts as a factor in production, i. e., an occupational activity in the broadest sense. The physiologic consideration of this activity means that it is primarily a physical activity. The distinction between physical and mental work is essentially arbitrary because mental work is linked to the body. With their methods, physiologists can hardly do justice to mental work; thus, this is primarily the task of work psychology.

Work physiology has the especially important social task of humanizing the working place. In the first half of this century, the problem of heavy physical work was major. It was a question of reducing stress by better adapting the machine to man but without a decrease in performance, thus avoiding "wear-and-tear" diseases. Today, the psychophysical problems predominate.

At least in the industrialized countries, heavy physical work has only a minor role and some of the problems of work physiology have been displaced to the psychophysical area. Those which should be mentioned include the problems of monotony during assembly line work, psychic stress and work under time pressures. In many areas of the modern working world, there are excessive psychic demands along with reduced physical demands, a combination which favors the development of cardiovascular diseases. Within the realm of work science, a greater scope of cooperation with psychologists and engineers is applicable. Due to the reduced physical requirement or localized, repetitive requirements of the job, work physiology and sport physiology are developing well today.

There is a smooth transition to the area of occupational or industrial hygiene. In some countries (e. g., Japan), the problems of applied work physiology are treated by occupational hygiene, which is primarily concerned with environmental influences on working man, the effect of dust and gases, nutrition and housing, daily routines and social life. For the most part, these questions concern work physiology as well as occupational hygiene.

Work physiology is concerned with healthy working man. The problems associates with the reciprocal actions between sickness and occupational work are treated in the area of occupational medicine.

In this connection, rehabilitation research should also be mentioned because it has the task of returning injured workers to occupational life.

B. Work Physiology and Technology

The working place and the worker represent a functional unit which operates optimally only when the worker is adapted to his task and the task is adapted to the worker. Numerous basic physiologic facts must be heeded if this adaptation is to be optimal. The worker must be trained and must practice for his task. However, experience indicates that practice and training have narrow limits. The manufacturer of a machine can generally adapt his machine to the physiology of man when he has learned why it is important. The wide variability of occupational jobs requires a greater deliberation in each individual case. It often happens that the particular construction characteristics come into flagrant contradiction with physiologic requirements. An example from the area of transportation is the pedal arrangement in a motor vehicle with automatic transmission. In this case, the left foot no longer has any function whatsoever. It would therefore be much more practical to coordinate the right foot with the gas pedal and the left foot with the brake. In this way, the time required to move from the gas pedal to the brake pedal at the moment of danger would be eliminated. An alteration of the "historical" pedal arrangement is opposed, however, because each driver formed a conditioned reaction to the arrangement when he learned how to drive. In order to brake with the new arrangement, his left leg would have to learn a new conditioned reflex. As a result, the driver instinctively resists this improvement because there would be a greater danger during the time of the transition; this would later be reduced. In addition, the previously learned reaction would again predominate during dangerous moments.

This small example may clarify the difficulty which often opposes an adaptation of the machine to man. The relationships are also complicated because the factors have opposing influences. At the same impact, for example, the recoil of a sledge hammer is related to its weight. To reduce the injuries from the recoil, the weight of the sledge hammer must be increased. There are limits to what can be done because the sledge hammer also must be held. When the weight is increased, the muscles which hold it will therefore fatigue more; as a result, these muscles can no longer compensate as well for the impact. This example shows that not only must the machine be adapted to man but the relative optimum between many negative influences must also be found.

C. Work Physiology and Economics

In recent years, the proportion of the population in industrialized countries which must perform heavy physical work has been reduced. For the most part, machines are used in their place. In the developing lands, heavy physical work must still be performed because the provision of machines would require capital investments which are not possible. Roughly estimated, energy supplied to the body via mixed nutrition costs 100 times more than energy from oil or coal. A country the size of India (approximately 600 million inhabitants) has perhaps 100 million inhabitants who daily perform heavy muscular exercise. Assuming for each of these 100 million inhabitants that only 1,000 kcal/day of metabolism can be spared by the conversion to machines, then \$100 million would be spared per working day if a person had to pay a price of \$1 for each 1,000 kcal of food. With 300 working days, an economy on a national scale of about \$30 billion can be calculated.

Such a gross estimation should make it clear that work physiology and the economy are closely linked to one another. Because man is still the most important production factor, at least for the entire earth, it is the task of work physiology to provide man with optimal working conditions and to place machines where heavy physical work must still be performed. Man's strength as a production factor is not primarily related to his muscles but to his intelligence to make instruments useful in the widest sense.

Sport Physiology

Sport and physical work have in common the fact that man uses his muscles under both conditions and sets into activity the entire sequence of reactions in the body which are associated with muscle activity. Physiologically, both forms of activity generally have only a quantitative but overlapping difference which is primarily associated with the level of intensity. The world of sports is as multifaceted faceted as the world of work, so that only basic principles can be elaborated in a comprehensive discussion of sport physiology. The load placed on organs, information channels and the cerebral data processing system during motor sports in a modern racing car, for example, is diametrically opposed to that operating on the marathon runner. Therefore, only the general physiologic operational patterns can be considered from the viewpoint of elevated work loads or those work loads at the limit of physical work capacity. This is the essential difference between sport physiology and work physiology; during work, this limit is most probably never reached.

Even though sport is occupational work for all professional athletes and serves as a way of earning a living, further differences between occupational work and athletic activity exist. Naive souls might believe that it is completely absurd to diminish the energetic requirement or to abolish heavy physical work with large financial expenditures and then discover that man has problems due to physical inactivity. As a result, the same financial expenditure is needed to construct sports facilities and to propagate sport for health reasons. The difference between the two forms of physical activity is less in the physiologic than in the psychologic realm. It is after all not an indifferent matter whether a performance is carried out with pleasure or reluctantly, even though a black-white distinction is naturally contained in this statement. Furthermore, environmental-hygienic and sociopsychologic differences must be considered. It would go beyond the bounds of this book to discuss the details here.

Apart from the necessary fundamental research, the tasks of sport physiology vary in the different fields of sport. In mass sport, which is the realm of sports for everyone, it must certainly take into consideration the health aspects for all age groups. Here it is a question of combatting physical inactivity, with its detrimental effects on the cardiovascular system, through early motivation for a type of sport which an individual can perform his entire life (life-time sports), such as swimming, skiing or tennis, i.e., those types of sports which are energetically demanding but not too strenuous. It is essential in this case to find the correct level and to give appropriate guidance to sportsmen and their trainers. In this regard, the Roux rule is valid: small stimuli are useless, moderate stimuli can be used, large stimuli injure. Such attempts are possible only in association with the usual sport sciences and with practice.

Performance sports and high-performance sports create entirely different problems. In these, active athletes and their trainers are especially interested in a verification of the results of their training methods in the physiologic sense. The physiologist can obtain numerous measurements which reflect a highly reliable indication of the athlete's condition. Of course, the influence of sports medical activities on effective athletic performance should not be overestimated. It can only be one factor in a series of sports scientific and practical measures, such as having good models as parents, preschool and school sport education and good coaches.

The sport physiologist should also have the courage to warn the athlete in those instances where high performance is deteriorating. This is always the case when injuries occur due to overtraining and especially when this occurs in children who do not have the necessary supervision. For an adult, high-performance athlete who is at his well-recognized limits of injury, athletic success may bring him compensation, enjoyment and social prestige; he thus takes all of these factors into his calculations. Personally, I find it intolerable when children are pushed through the water for several hours daily in order to win swimming competitions for the club, for the reputation of the country or for the "better system."

Up until now, training research has been predominantly limited to the systems whose performance has been improved. In the future, we will have to work harder to examine the limits where disadvantages or even injuries are to be expected due to inordinate amounts of training. In this regard, it has already been shown that the quality of blood pressure regulation at rest is worsened by endurance training. A similar situation can be found for the sensitivity of the respiratory system to different stimuli (see p 295).

Within the framework of our discussion, it is obvious that we cannot present the physiology of special types of sport in detail but only the general basic principles. However, these will be elaborated , such that the limits of performance of the stressed organs and organ systems will be made clear, and the athlete, the teacher or the coach can see the valid relationships and limits for his own type of sport from these data.

1 Muscle Activity

To react in any way, the body requires muscles because they alone give it the possibility to perform, whether it is the production of work or the relaying of information in the form of speech or a gesture. Of primary interest is voluntary muscular activity produced by the skeletal musculature (striated muscle) which, in contrast to smooth (involuntary) muscle and cardiac muscle, is able to rapidly contract and relax.

1.1 Structure of Muscle

The structural elements of skeletal muscles are the muscle fibers which have an average diameter of 50–100 μm and a length of up to 10 cm (100,000 μm) and more. On the outside, they are enveloped by a typical cell membrane, the sarcolemma. On the inside, in addition to their cytoplasm (sarcoplasm) and numerous nuclei (up to several hundred), they contain even smaller, threadlike structures running longitudinally, the myofibrils (average diameter 1 μm). Under a light microscope, these exhibit a regular sequence of alternating dark and light zones, which in their entirety give the skeletal muscle fiber its cross-striations. The dark zones are labeled A bands and the light zones I bands (Fig 1). Each of these bands is additionally transected by a narrow line; the I band by the darker, striped Z line, the A band by a lighter H line (Fig 6). The structural unit of each myofibril is the sarcomere, which is the space between two Z lines.

Under the electron microscope, it can be seen that the myofibrils are made up of even finer, threadlike subunits, the myofilaments (Fig 2). These can be differentiated into two types based on thickness and chemical composition. The one has a diameter of 5 nm (50 Å), is only one third as thick as the other and is composed of a complicated protein molecule, actin. It is thus labeled the actin or "thin" filament. The "thick" filament (diameter 15 nm or 150 Å) is made up of the protein molecule myosin and extends over the entire A band. The actin filament originates at the Z line and extends to the H lines. Thus, they extend a considerable distance between the myosin filaments and into the A band. The bands recognizable under the light microscope are thus caused by the regular alternation of differing thicker and threadlike protein molecules (filaments) which are also organized in a close geometric fashion relative to one another. Because two thin filaments are interspersed on both sides of two neighboring myosin

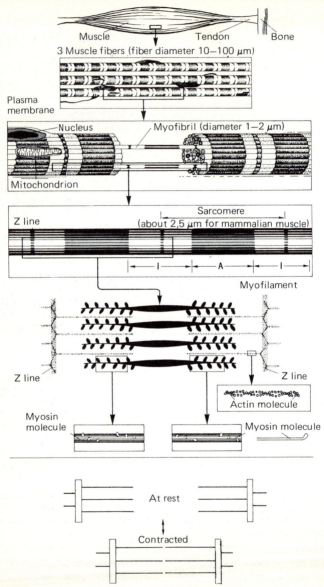

Fig 1   Gross and fine structure of muscle (schematic). From Novikoff, A. B. and Holtzmann, E.: Cells and Organelles (New York: Holt, Rinehart & Winston, 1970).

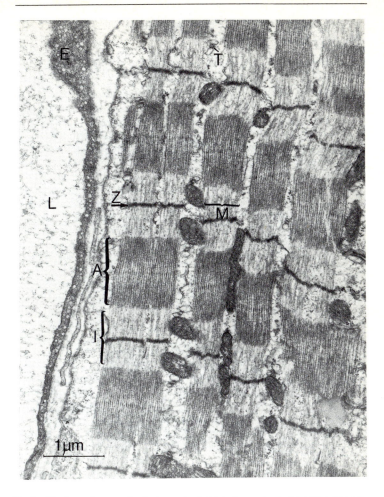

Fig 2 Electronmicroscopic photograph of striated muscle. A, A band; I, I band; M, Myofibril; T, Tubule; Z, Z line; E, Capillary endothelium; L, Lumen. Courtesy of Prof. Dr. Hammersen, Institute of Experimental Morphology, Deutsche Sporthochschule, Cologne, FRG.

filaments, on the average there are 6 actin filaments surrounding a centrally located myosin filament. During contraction, there is formation of an actomyosin complex, resulting in reversible cross-bridges between the two filament sites which simultaneously move against one another. The thin actin filaments progressively glide between the thick myosin filaments and the I band eventually can completely disappear (bottom of Fig 1).

1.1.1 Mitochondria and Sarcoplasm

The mitochondria are the sites of oxidative energy production. In the mitochondria, oxygen reacts with the hydrogen extracted from nutrients by a dehydrogenation enzyme (p 49). Energy-rich phosphate bonds are formed which can then be used by the organism for energy.

Mitochondria are filament-like or granular portions of the cell constructed with a double membrane made of lipoprotein. The inside wall is articulated with many protrusions (cristae). The enzymes are on the surface of these structures. The most important morphological characteristic is a thick border with spherical particles; these can be isolated as mitochondrial adenosine triphosphatase (ATPase). These particles are combined with the membrane due to a coupling factor. It is probable that during oxidative energy production, inorganic phosphate is transformed into adenosine diphosphate (ADP) in the mitochondrial ATPase (p 45). In the nutrient exchange between the cytoplasm and the mitochondrial space, there are specific transport systems for metabolites of the citric acid cycle (p 47), phosphate, ATP and ADP. A smooth, external, highly-permeable membrane envelops the inner membrane convolutions.

The mitochondria of smooth muscle are small and situated near the cell nucleus. In the striated muscle discussed here, they are found in the interfibrillar clefts. In phasic muscle fibers, they are small and irregularly placed; in tonic musculature, they lie coupled on both sides of the Z lines. In the continuously active musculature of the heart and diaphragm, they are densely located in the interfibrillar clefts.

All enzymes of anaerobic energy production (p 45) form relatively small molecular particles which are soluble in the sarcoplasm. They are a major component of the sarcoplasm. In the cells of smooth muscle, these enzymes are distributed regularly throughout the myoplasm. In striated muscle, together with particles of glycogen, they are almost exclusively located in the isotropic sections.

1.2 General Fundamentals of Excitation

1.2.1 Resting Membrane Potential

Each cell in the human body is enveloped with a functional membrane which is selectively permeable, i. e., not penetrable by all ions in the same manner. In the cell are primarily potassium ions and large negative anions, which are special protein ions. Outside the cell in the interstitium are primarily Na and Cl ions, plus smaller concentrations

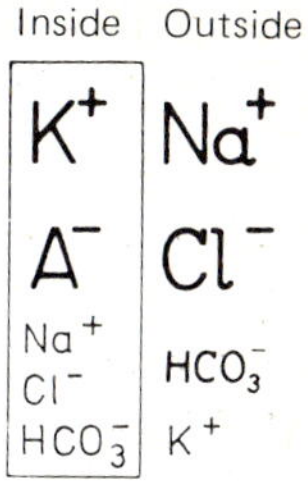

Inside   Outside

$$K^+ \quad Na^+$$

$$A^- \quad Cl^-$$

Inside	Outside
Na^+	HCO_3^-
Cl^-	
HCO_3^-	K^+

Fig 3   Concentration distribution of various ions within and outside the cell. Size of letters indicates relative ion concentration. From Schneider, M.: Einführung in die Physiologie des Menschen, 15th ed. (Berlin: Springer, 1966).

of HCO_3 ions. The distribution of ions is schematically represented in Figure 3; the size of the letters indicates the relative ion concentrations inside and outside the cell. However, such a concentration difference could only persist if it were present from the beginning and if the cell membrane were completely impermeable to every type of ion. This is not the case because small amounts of ions, especially potassium, are continually diffusing across the membrane in relation to their concentration gradient. In a given time, dependent on membrane permeability, the concentration differences would be completely equalized. The maintenance of the concentration differences is an active achievement of the organism. If the tissue oxygen supply is interrupted for a long period of time, then the actual ion concentration difference becomes zero. The concentration difference is maintained by an active metabolic process. Therefore, there is an "ion pump function," by which the Na^+ ions are pumped out of the cell by the sodium pump and the K^+ ions are pumped into the cell by the potassium pump. In all unstimulated cells, the permeability for K^+ ions and Cl^- ions is the greatest. In the case of Na^+ ions, the permeability is about 10–25 times less than that for K^+ ions under resting conditions. The Na^+ ions which enter the cell are immediately pumped to the outside, so that the membrane is considered functionally impermeable. Protein anions cannot diffuse due to their size.

Parallel to this ion pattern is a resulting difference in electric tension (potential) between the interior and surface of the cell. This potential is called the resting membrane potential, which is essentially a potassium diffusion potential for the following reasons. The anions to the potassium ions are the large, impermeable protein ions, which are negatively charged. As a result of their permeability, the potassium ions tend to diffuse toward the outside of the cell because their cell concentration is much higher. However, they are held back because of the anions. The result is a tension, as illustrated and clarified in Figure 4. A second tension, which is comparatively small, is generated by the chloride ions, whose anions are the sodium ions located outside the cell. The membrane potential is about –50 to –100 mV, depending on the type of cell and the ion distribution. It is possible to measure this

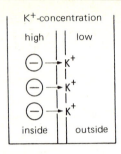 

Fig 4 Schematic explanation of resting membrane potential as potassium diffusion potential. The cell membrane which separates interior and exterior is permeable to potassium ions on the left. These ions attempt to move to the right (exterior) in relation to the concentration gradient. Anions cannot follow them through the membrane due to their size and so they hold potassium ions inside the cell because of their opposing electric charge. This tendency toward separation expresses itself as potential difference which holds them in equilibrium. As a result, there is a positive charge on the edge of the membrane bordering the solution with a lower potassium concentration. Picture on right demonstrates analogous development of tension. From Lullies, H.: In: Kurzgefasstes Lehrbuch der Physiologie, 2nd ed., ed. by Keidel, W. D. (Stuttgart: Thieme, 1970).

potential if a microelectrode is inserted inside the cell and another electrode is applied to the surface of the cell.

1.2.2 Local Excitation

Those cells which have the specialized property of excitability can be stimulated; this is especially true for muscle cells, nerve cells and receptor cells. This excitation occurs due to a mechanical or chemical stimulation. The details of how this happens are still largely unknown.

Under the influence of a stimulus, the membrane becomes permeable to the Na^+ ions, which then penetrate into the interior of the cell due to their concentration gradient. Because there are now numerous positive charge carriers which have come into the cell, the resting potential becomes less and the cell is "depolarized." When the stimulus affects the cell for only a short time, then the level of excitation returns to the normal resting value. This initial depolarization is also called "local excitation" and as such, remains under the threshold value.

1.2.3 Transmitted Excitation

If the depolarization surpasses a certain value (threshold potential), there will be a transmitted or propagated excitation. Due to an influx of even more Na$^+$ ions, the cell interior briefly becomes positive relative to the outside of the cell. As a result of this shift in potential, there will now be a massive outflow of K$^+$ ions. During the excitation

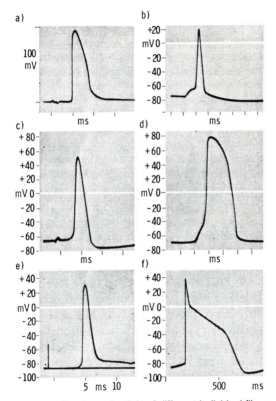

Fig 5 Resting and action potentials of different individual fibers and cells. a, node of Ranvier of frog motor nerve fiber; b, motor anterior horn cell of the spinal cord of the cat; c, giant axon of the cuttlefish; d, electroplate of electric organ of electric eel; e, frog sartorius muscle fiber; f, Purkinje fiber from sheep heart. The curves, which come from different investigations, have all been reduced to the same ordinate scale. Potential differences were recorded with inserted microelectrodes, with the exception of a. All graphs show resting potentials of about the same value of –80 mV and a similar pattern of action potential rise. The "overshoot" into the positive zone is particularly large in the electroplate of the electric eel. The biggest differences can be found in the decline of the action potential. Observe the larger time scales of e and f and the multiphasic discharge of the repolarization in the Purkinje fiber; this is also obvious in the node of Ranvier. From Lullies, H.: Kurzgefasstes Lehrbuch der Physiologie, 4th ed., ed. by Keidel, D. (Stuttgart: Thieme, 1975).

process, a characteristic pattern of changes in potential is seen. Starting from a resting potential of about –80 mV, depolarization begins with an influx of sodium, i. e., there is a sudden change in tension toward positive values which can reach +20 to +30 mV. Immediately thereafter, a repolarization is created by two reactions. First, the sodium influx is halted, causing the membrane to become impermeable to sodium once more. Next, the outflow of potassium causes a return in the direction of the resting potential. The basic pattern of such an action potential is shown in Figure 5. Although the pattern is somewhat different from one type of cell to another, there is always a rapid depolarization phase and a rapid and slow repolarization phase.

The transmitted excitation thus arrives at the state where the accompanying local action potential current of the cell membrane stimulates the unexcited adjacent section of the membrane. This stimulates the next adjacent section in turn, so that the excitation radiates out like a ripple (wave theory). After the excitation wave has run off, the ion pumps once more take over to restore the ionic equilibrium. For a short period of time after the excitation, the membrane is refractory, i. e., no new excitation can be generated.

1.3 Excitation and Contraction of the Muscle Fiber

The excitatory mechanism of the muscle fiber follows the same fundamentals as are valid for general excitation. These fundamental processes occur at the membrane separating the sarcoplasm from the extracellular fluid. Here also there is a selectively permeable membrane which is not uniformly permeable to all ions. Normally, each contraction of skeletal muscle is initiated by an excitation. However, under experimental or pathologic conditions, both processes can occur independently. Excitation of the skeletal muscle fiber is caused by the transmitter acetylcholine which is secreted by the motor end-plate. The motor end-plate is a part of the synapse (p 24) and represents the functional connection between the neuron and the skeletal muscle fiber. The tissue hormone acetylcholine increases the permeability of the muscle cell for Na^+. In the muscle fiber itself, there is an enzyme cholinesterase, which can rapidly split acetylcholine into two biologically inactive products, choline and acetic acid. At any given moment, the effective acetylcholine concentration is dependent on the amount synthesized per unit of time and the amount degraded per unit of time. At rest, the membrane potential of skeletal muscle fibers is about –80 to –90 mV. The greater the acetylcholine concentration, the greater its effect and the stronger the depolarization of the membrane. As a result, the membrane potential is measurably reduced. As previously described, this condition is called the local response. When the influx of Na^+ ions becomes so great that the membrane potential is reduced to less than –60 mV, a transmitted excitation is released; this spreads out in both directions as a wave of excitation from the point of origin, the end-plate, toward the ends of the muscle fiber at a velocity of about 1 m/s. Approximately 0.001 s after the depolarization has occurred, the permeability decreases. The membrane then becomes repolarized, so that it again becomes receptive to a new stimulus. Because those excitations which are transmitted must also be maximal and surpass the threshold value of the membrane potential, muscle fiber membranes are one of the structures which follow the all-or-none law. The excitation wave can be displayed as the action potential.

1.4 Fundamentals of Electromechanical Coupling

How is the electric stimulation now transformed into a mechanical contraction? During a contraction, both specific muscle proteins, actin and myosin, form the compound actomyosin when the energy-rich ATP is broken down into ADP and inorganic phosphate (P_i). This happens only when free calcium ions are also present in a specific concentration in the sarcoplasm. In resting muscle, this ion is present only in the intrasarcoplasmic pore spaces bordering the membrane. These cavities surround the myofibrils in the form of longitudinal tubules and are transversely connected with each other at definite intervals (Fig 6). In their totality, they are termed the longitudinal or L system and in the area of their most widely expanded cross-connections (terminal cisternae or lateral sacs), they come into contact at the elongated transverse or T tubules (diameter 50 nm), which always run across the muscle fiber (Fig 6).

The T tubules make deep invaginations in the sarcolemma. As a result, they are in continuous open association with the surface of the fiber and there is an interchange with the extracellular space (Fig 6). Frequently, a T tubule will be flanked on both sides by the parallel lateral sacs of the L system, with whose walls they are in direct contact (Fig 6). If an excitation wave travels over the sarcolemma of a muscle fiber, it is transmitted deep into the muscle along the membranes of the T tubules. This depolarization leads in a still unknown manner to a "state of permeability" in the membranes of the lateral sacs closely attached to the T tubules, from which calcium ions can now pass into the sarcoplasm. In this way, the critical threshold concentration for the initiation of the interaction between myosin and actin is exceeded, so that with the assistance of ATP, an initially small number of actomyosin compounds can be formed.

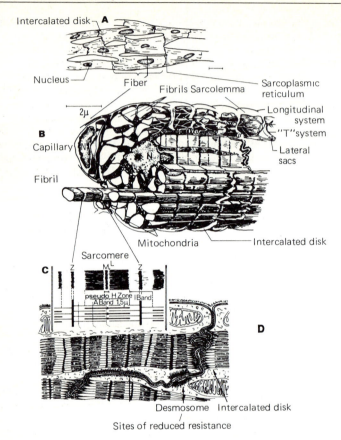

Fig 6 Summary of structure of myocardium. A, syncytial arrangement of myocardial fibers. B, construction of muscle fiber from fibrils, as well as pattern of sarcoplasmic reticulum and blood supply; C, schematic structure of fibrils; D, electron microscopic representation of fibrillar structure, as well as sites of reduced resistance ("tight junctions") and junctions of high mechanical tensile strength ("intercalated disks") between individual muscle fibers. From Ganong, W. F.: Medizinische Physiologie (Berlin: Springer, 1971).

This actomyosin itself has the associated effect of an ATP-splitting enzyme (ATPase). It thus causes a chain reaction. The more actomyosin is present, the more ATP is split and the more energy is produced for the contraction. This process is controlled by the free calcium ions in the sarcoplasm. As soon as no more free Ca^{++} ions are present in the sarcoplasm, there is a drop in the ATP-splitting activity. Via the oxidative metabolism, the ADP + P_i is once again regenerated into ATP. As a result, actomyosin is split again into myosin and actin and the muscle fiber relaxes.

The Ca^{++} concentration in the sarcoplasm depends on two factors: the amount per unit of time liberated from the lateral sacs and the amount pumped back into the sacs per unit of time. The liberated amount depends naturally on the value per time of the excitation wave traveling over the fiber, whereas the amount pumped back is dependent on the capacity of the calcium pump. Therefore, the more excitation waves per time which travel over the fiber, the stronger the muscle will contract. The participating metabolic processes related to contraction will be discussed on page 44. In the case of smooth muscle fibers, calcium penetrates the cell from the outside.

1.5 Mechanical Characteristics of Muscle Fibers

1.5.1 Reactions to Passive Stretching

In the experimental setup of Figure 7, when the spiral spring is stretched with a spring constant E by the amount ΔL, then the following force F is applied:

$$F = E \cdot \Delta L \quad \text{(Hooke's Law)} \tag{1}$$

This relationship is illustrated in Figure 8 for springs with differing spring constants E. If the spring is stretched further, the relationship between force and extension remains the same at each point on this line. That is, the spring is ideally elastic and follows Hooke's Law. Elasticity is appropriately defined as the capacity of a material to store reversible deformative forces.

Skeletal muscle also exhibits an elastic reaction, only its extensibility does not obey Hooke's Law. The force necessary to stretch a muscle increases disproportionately to the extension, as schematically illus-

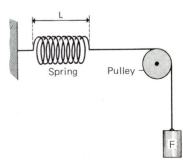

Fig 7 Experimental setup for determination of relationship between applied force tension and length of a spring (Hooke's Law).

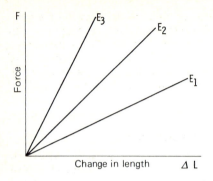

Fig 8
Change in length is proportional to force applied. Extent of change in length is dependent on spring constant E.

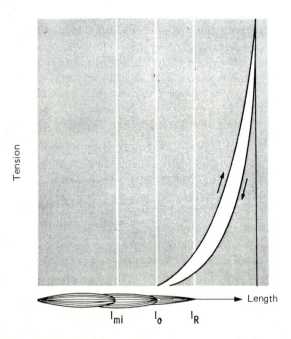

Fig 9   Relationship between length and tension during passive stretching of a muscle; mi, minimal length; o, length at onset of tension; R, resting length of muscle in the body. It can be observed that the muscle does not obey Hooke's Law. Part of the tension is lost between processes of stretching and recoil due to plastic deformation. Lost tension is represented by white area. From Brecht, K.: In: Kurzgefasstes Lehrbuch der Physiologie, 4th ed., ed. by Keidel, W. D. (Stuttgart: Thieme, 1975).

trated in Figure 9. In this case, the muscle no longer shows the same relationship between force and length during the recoil after stretching, as was found during the process of stretching. In contrast to the spiral spring, if muscle is passively stretched a portion of the work of extension is lost; this lost work is represented in the diagram by the white area. Muscle does not have a pure elastic behavior but exhibits an additional plastic reaction.

1.5.2 Reactions during Active Shortening

Basically, there exist two extreme situations in relation to contractions of muscle fibers, namely, isometric and isotonic contractions. An isometric form of contraction occurs when the muscle only develops force while maintaining a constant length and an isotonic contraction occurs when the force remains constant and only the length is changed. All forms of contraction lying between these two extremes are called auxotonic. Practically speaking, there are also mixed isotonic-isometric contractions. For example, when a weight is lifted, the result is an

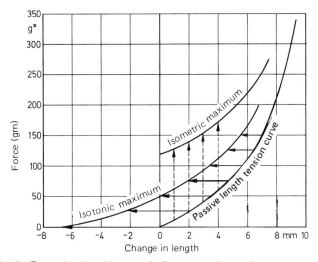

Fig 10   Force developed by muscle fiber under isometric contraction conditions and during shortening under isotonic contraction conditions. In both cases, force developed is dependent on degree of prestretching of the fiber. The greater the mechanical prestretching of the muscle fiber, the more force it can develop or the less the individual muscle fiber can shorten. Thus, for each contraction under similar patterns of excitation, the mechanical work performed is dependent on the prestretching of the muscle. From Reichel, H.: Ergebn. Physiol. 47: 469, 1952.

isometric contraction until the fiber force matches that of the weight. After that, there is an isotonic phase because now the weight remains constant but there is a change in length. This form of contraction is called an afterloaded contraction.

As we have already seen, under physiologic conditions the contraction of a skeletal muscle fiber is initiated by the motor end-plate. This end-plate normally sits in the middle of the muscle fiber. When an excitation of the end-plate and a transmission of the stimulation is initiated, the two excitation waves spread out from the middle of the muscle toward the ends, causing a contraction. The result is a measurable twitch, whose velocity under otherwise similar conditions is usually a constant. The force depends on the amount of prestretching of the fiber, as demonstrated in Figure 10. On the abscissa is the change in length, with the initial length of the muscle fiber given a value of zero.

The mechanical work performed during each twitch under the same excitation conditions is therefore dependent on the prestretching of the muscle. The amount of work is the greatest with moderate prestretching.

Although twitches have no importance in voluntary movements, their description illustrates more clearly the fundamental characteristics of skeletal muscle fibers. Voluntary movement results more often from multiple waves of excitation traveling over the membrane. The contraction responses then become superimposed and fuse into a state of tetanus, as schematically illustrated in Figure 11. If the motor nerve of

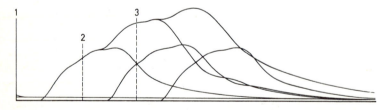

Fig 11   "Summation or superposition" of individual twitches of muscle during artificial stimulation. Three electric stimuli are imposed on the muscle. A vertical line (1, 2 and 3) is used to denote the moment of stimulation. After a corresponding latent period, each stimulus causes a characteristic muscle twitch curve, as shown in the 3 successive curves in the lower part of the diagram. If the 3 stimuli are applied in rapid succession, however, such that the 2nd stimulus appears before the muscle twitch from the 1st stimulus has died out, there the 2nd curve is superimposed on that of the 1st; the same thing occurs with the addition of the 3rd curve. The result is a temporarily extended and increased muscle contraction. From von Frey: In Schneider, Einführung in die Physiologie des Menschen (Berlin: Springer, 1966).

a muscle is stimulated above its threshold value at time T, after a latency period a double excitation wave is released from the motor end-plate toward both ends of the muscle fiber. This causes a contraction (twitch 1). If the nerve is stimulated once more at a time when the twitch has already discharged itself, then there will be a second twitch after a new latency period (twitch 2). If the second stimulus is moved back closer to the first one, e. g., when the first twitch has not been completely discharged, then a second excitation wave travels over the fiber and inhibits the muscle from prematurely relaxing. As a result, the fibers remain in a state of contraction. The closer the second stimulus is to the first, the smoother will be the oscillations in the curve and the greater will be the muscle force developed. The greater the number of excitation waves traversing the fiber at the same time, the greater will be the increase in strength and the longer the fiber contraction will be maintained. However, there is a limit to the number of excitation waves due to the refractory period; after each stimulation the skeletal muscle membrane is no longer excitable for at least 0.001 s.

The state of contraction is dependent on the concentration of Ca^{++} in the sarcoplasm. Each stimulation liberates a defined microamount of Ca^{++} from the lateral sacs, which is later pumped back. The greater the number of excitation waves per time flowing over the fiber (compared to the rate that Ca^{++} is being pumped back), the longer will be the duration and the greater will be the strength of the contraction. The functional principle in the organization of muscle is not represented by the individual fiber but by the "motor unit," which is defined as a motor anterior horn cell and the muscle fibers which it supplies. These motor units are not rigidly limited but can overlap one another. The relationship between the number of nerve fibers and the number of muscle fibers which they supply is called the innervation ratio. The larger this ratio is, the finer the movements and the gradations of muscle strength will be. As an example, the innervation ratio in muscles of the sensory organs (eye musculature, tensor tympani) is approximately 1 : 7, whereas that of the quadriceps femoris is 1 : 1,000. Each individual muscle fiber possesses a motor end-plate, which is ordinarily situated at its middle. There are also especially long muscles, such as the sartorius, which have several end-plates. The gradations of force developed by the total muscle can also be controlled by the number of innervated motor units. As a general rule, it appears that only about two thirds of the muscle fibers of a muscle can be voluntarily innervated at the same time. All muscle fibers can be made to contract simultaneously only involuntarily via a proprioceptor reflex. In this case, the force produced is close to the tensile strength of the muscle. As a result, under unfavorable conditions there may be a tearing of the muscle.

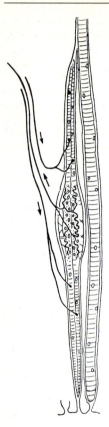

Fig 12   Schematic representation of muscle spindle. Spindle consists of an extension-sensitive sensor in the middle which sends information about tension to the spinal cord and two contractile sections which are stimulated by the γ fibers. From von Holst, E.: Fortschr. Zool. 10 : 352. 1956.

Depending on the situation, with almost all types of work (especially precision work) it is possible to exactly control strength and velocity. Normally, movement occurs tetanically, i. e., in the form of a prolonged contraction. In this instance, muscle strength is varied by altering the number of excitation waves in an individual fiber, as well as by the number of active motor units.

The force and velocity of muscle movement is controlled via the muscle spindles (p 26). These spindles are particularly numerous in the extensor muscles of the extremities and are somewhat less frequent in the flexors. In addition, they are especially dense in the muscles of the human hand and in those of the eyeball. They are also present in the tongue, larynx, diaphragm and tensor tympani. Figure 12 shows a simplified sketch of a muscle spindle. Note that it is not stretched between different muscle fibers, but runs parallel to the longitudinal axis of one fiber. The muscle fibers of the spindle itself are motorically innervated on both ends via thin nerve fibers, whose thickness is 3–10 μm (the so-called γ or gamma fibers). In the middle of the spindle, there is a heavily nucleated section (nuclear bag) enclosed within a capsule and lymphatic spaces, which is neither striated nor contractile. This nuclear bag is encased in many branches of at least one 15–17 μm sensory nerve fiber. Although the motor nerve fibers of spindles almost never have branches to the other muscle fibers, they do supply branches to other spindles. The remaining muscle mass is supplied by the thick nerve fibers of 10–20 μm diameter (the so-called α-fibers).

1.6 Fundamentals of Motor Activity

The spindle system is the "final consumer" of motor activity. To better understand motor activity, we must first take a closer look at its basic elements and their interconnections. Then we will return to the function of the spindles (p 26).

1.6.1 Structure and Function of Neurons

A neuron is defined as a nerve cell with all its anatomical processes; a motor neuron is illustrated in Figure 13. The nerve cell contains numerous short projections (dendrites); most have only one long projection (axon), whose length can be as long as 1 m. These projections make it possible for the neuron to have connections not only with other neurons, but also with muscle fibers, glands and other organs. The axons represent the motor nerve fibers of the neurons.

Two types of nerve fibers are distinguished, myelinated and nonmyelinated. The axis cylinder of the axon is surrounded by an axon

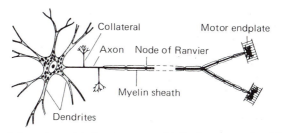

Fig 13   Schematic model of motor neuron. Multipolar nerve cells of the spinal cord with dendrites and myelinated axons distribute themselves in periphery and supply numerous muscle fibers. From Lullies, H.: Erregung und Erregungsleitung; Nervenphysiologie, In: Kurzgefasstes Lehrbuch der Physiologie, 4th ed., ed. by Keidel, W. D. (Stuttgart: Thieme, 1975).

membrane whose thickness is about 8 nm. In the case of myelinated nerve fibers, there is a myelinated sheath on the outside of the axon membrane to separate it from the neurolemma. This sheath is interupted at intervals of 1–3 mm by constrictions; these are called the nodes of Ranvier.

The diameter of individual nerve fibers is variable. Myelinated fibers vary between 3 and 10 µm and nonmyelinated fibers have a diameter of only 1–2 µm. Nerve conduction velocity is somewhat proportional to fiber thickness. Fibers between 10 and 20 µm are termed α fibers, between 7 and 15 µm β fibers, between 4 and 8 µm γ fibers and between 2.5 and 5 µm δ fibers.

Axons can send off collaterals to other axons.

Nerve conduction in nonmyelinated fibers follows the principles of propagated excitation already discussed, by which the resulting ion flow depolarizes the adjacent part of the membrane. However, in myelinated nerve fibers, there is a different mechanism, i. e., there is a "saltatory" transmission. The stimulus is amplified in the nodes of Ranvier. It then jumps from node to node, resulting in a more rapid conduction velocity.

1.6.2 Structure and Function of Synapses

The junctions between two neurons, from a neuron to a muscle fiber or to other effectors, are called synapses. With the arrival of an action potential, a transmitter substance is liberated at the terminal branches of the axon. This transmitter is one of the tissue hormones liberated by nerve cells; it passes through the terminal membrane, diffuses across a narrow gap filled with fluid and acts on the membrane of the adjoining cell. The detailed construction of a synapse is presented in Figure 14. As can be seen in this diagram, the supplying axon branches out into many nonmyelinated fibers. These are finally transformed into knob-shaped terminal formations, in which numerous mitochondria and fine, fluid-filled vesicles are found. It is these vesicles which produce and store the transmitter substance. This transmitter crosses over the synaptic cleft and acts on the postsynaptic membrane belonging to the adjacent cell, producing an excitation of the postsynaptic membrane. The synaptic excitation is basically different from the action potential excitation transmitted by the axon in that it is not transmitted over a long distance, i. e., it is locally limited to the region of the synapse.

There are two elementary responses to synaptic stimulation; these are excitatory and inhibitory. The excitatory response occurs at synapses which secrete a depolarizing transmitter. An inhibitory response appears when the transmitter hyperpolarizes the membrane, i. e., it increases the membrane potential. Thus, a depolarizing transmitter

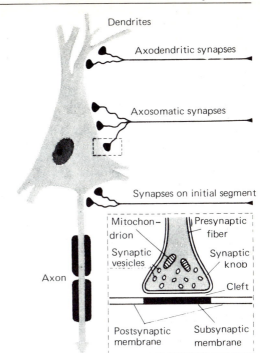

Fig 14
Diagram of different synaptic contacts between two neurons. Bordered part shows structure of synapse under stronger magnification. From Caspers, H.: Zentralnervensystem, In: Kurzgefasstes Lehrbuch der Physiologie, 4th ed., ed. by Keidel, W. D. (Stuttgart: Thieme, 1975).

lowers the threshold for the transmitted excitation and a hyperpolarizing one raises it.

The chemical structure of the transmitter substances is only partially known. The transmitter substance at some synapses is either adrenalin (epinephrine) or norepinephrine; it is acetylcholine at others. Amino acids also have a decisive role as the transmitter substance at the synapses of the central nervous system (CNS).

The fact that the transmitter vesicles are always found at the terminal points of the axons means that the synapse can conduct its excitations in only one direction, even though the nerve itself is capable of transmitting stimuli in both directions.

1.6.3 Inhibition, Facilitation and Summation in the Neural Network

In the section "General Fundamentals of Excitation," it was determined that an excitation is initiated by a stimulus and that there is a selectively increased permeability of the stimulated membrane to sodium ions. The local excitation becomes a transmitted stimulus only when the depolarization surpasses the threshold value. The stimulus causes a synapse, producing the effect of the transmitter on the postsynaptic membrane. In this case as well, there is a local excitation which will be transmitted if the strength of the depolarizing stimulus is large enough to surpass the specific membrane threshold. These general membrane characteristics, together with the fact that the transmitter will be nullified by the appropriate enzyme, makes the synapse a sort of "computer" capable of spatial and temporal summation.

To clarify the processes of the synapses, take the example of a cholinergic synapse, whose transmitter substance is acetylcholine. With each action potential beginning at the axon and arriving at the synapse, a fixed microamount of acetylcholine will be released. This, in turn, is affected by the enzyme cholinesterase and is broken down into the inactive components of choline and acetic acid.

This process requires time, however. The effective concentration of acetylcholine is a function of the number of action potentials per unit of time and the degradation rate of the cholinesterase. According to the law of mass action, the higher the concentration of acetylcholine, the faster will be the degradation. If the incoming action current frequency is too low, then the critical concentration needed for the initiation of a transmitted excitation to the postsynaptic membrane will not be attained. If the frequency is higher, the different action potentials will be temporally summated and the excitation will appear when the liberation first surpasses the rate of degradation. If the neighboring ends of two axons are active, then a fixed amount of transmitter will be liberated from each of them; these will then be spatially summated.

The threshold of the postsynaptic membrane is modified with additional hyperpolarizing or depolarizing impulses. If there is more hyperpolarizing substance (inhibition), then the critical concentration of the transmitter has to increase in order to transmit the excitation to the postsynaptic membrane. On the other hand, there will be a reduction (facilitation) with the entrance of more depolarizing substance. The advantage of such a system is obvious. The synapse does not transfer only simple action potentials, but is in the position to appropriately modify the inflow:outflow ratio between two neurons via the additional influences of other neurons. The synapse can be compared with the operation of hybrid computers which are used in technology.

1.6.4 Structure and Function of the Muscle Spindle System (Monosynaptic Reflex)

The simplest form of controlled motor activity is represented by the monosynaptic (proprioceptor) reflex. A reflex is basically defined as

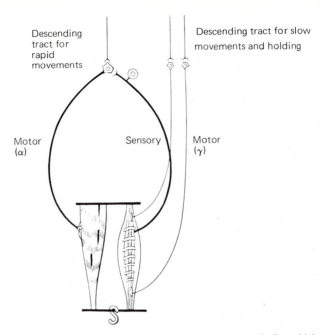

Descending tract for rapid movements

Descending tract for slow movements and holding

Motor (α)

Sensory

Motor (γ)

Fig 15   Schematic review of the function of the muscle spindle, which is arranged parallel to the working muscle. If the muscle is stretched, then the spindle is also stretched. As a result, the stretch receptors are stimulated and conduct the sensory stimulation to the synapse, where the α motoneuron is stimulated, initiating a contraction. This continues until the former tension level of the spindle is almost completely restored (holding control). The prestretching of the muscle spindle occurs via the γ motoneuron through a contraction of the intrafusal fibers, producing the set point (servomechanism). Rapid adjustments occur in an uncontrolled manner via a direct effect on the α motoneuron. From Schneider, M.: Einführung in die Physiologie des Menschen (Berlin: Springer, 1971).

an involuntary response to a stimulus. The reflex runs over a reflex arc, which is composed of a receptor, an afferent conductor, a switchboard, an efferent conductor and an effector. Most laymen today are aware of the testing method for the most widely known reflex in this group, the patellar tendon reflex. The thigh is supported to that lower leg can swing freely. Then, using either the edge of the hand or a small hammer, the patellar tendon is hit. The quadriceps femoris becomes stretched, it responds with a twitch and the lower leg swings up briefly. In healthy persons, it is possible to initiate similar reflexes with all of the skeletal muscles. The important thing is that the muscle be briefly stretched. The response is then always a twitch. The initiated mono-

synaptic reflex is the characteristic response of a control loop responsible for adjustments of the muscle length. It is called a monosynaptic
reflex because it travels over only one synapse in the CNS, i. e., it is
directly transferred from a sensory afferent nerve to a motor anterior
horn cell (α motoneuron). Figure 15 illustrates a control loop; an
introduction to the basic concepts of control can be found in the
supplement (p 301).

As already mentioned, a control loop has the task of maintaining the
length of a muscle at a predetermined value. This is independent of the
effects of any external forces on the muscle.

Each control loop must have a sensor which continuously measures the
control variable. This is the task of the middle, extensible portion of
the spindle, which is functionally suspended in the intrafusal spindle
muscle fibers at the origin and insertion of the muscle. When the
spindle is stretched, the action current frequency increases in the
afferent α fibers. This now activitates the α motoneuron via the
synapse. Thus, the real amplifier of the control system is in this area.
The α motoneuron itself guides only the effector, i. e., the extrafusal
fibers of skeletal muscle via the efferent α motor nerve fibers and via
the motor end-plate. Because the effector and receptor have the same
point of suspension, the spindle will be under tension until the initial
muscle length is approximately restored.

As long as the contractile state of the intrafusal fibers is constant, we
have an example of a true homeostat which, in contrast to the normal
resting tension conditions of muscle, holds the muscle length at a
predetermined value by means of active contractions.

The control loop can do still more, however. The intrafusal fibers can
modify their tension and the prestretching of the spindle. They are
regulated by the γ motoneurons of the spinal cord, which the intrafusal
fibers supply via the γ nerve fibers. A slow, controlled movement
results when the γ motoneurons are activated from higher levels of the
CNS. As a result of this activation, the intrafusal fibers contract and
stretch the spindle, which then begins to send increased numbers of
action potentials to the α motoneurons. The extrafusal muscle fibers
stay contracted until the spindle is practically relaxed once more.
Theoretically, this is a type of servomechanism, whose reference input
is regulated by the γ motoneurons.

In the previously mentioned test of the proprioceptor reflex, the
spindle was stretched briefly. As a consequence of the duration of the
stimulus transmission, the contraction first occurs only after the
stimulus has already faded away. Precise analyses of the reflex events
have shown that muscle spindle firing is not only proportional to the
amount of stretching but also is related to the velocity of stretching;
mathematically, this corresponds to the differential quotient of stretching. Thus, from a theoretical regulatory viewpoint, we are talking

about a proportional-differential or PD sensor (Suppl, p 304). The D portion is small in tonic muscles but large in phasic muscles.

From these relationships, the physiological significance of proprioceptor reflexes can be obtained. They contribute to the support of the body and its extremities. For example, if the gastrocnemius muscle is stretched by passive displacement of the body's center of gravity, then it will involuntarily contract to oppose the tendency to fall. Depending on the PD sensitivity of the receptors, the faster the force is imposed, the faster and earlier the counterreaction sets in. Even in the case of voluntary movements, the proprioceptor reflexes are in continuous association with the total motor activity, producing a smooth movement pattern which is essentially independent of external force.

The left side (agonist) of Figure 16 indicates that the described control loop contains even more elements. Tension sensors are located in the tendons (Golgi tendon organs) which will fire when the tendon is severely stretched. The afferent fiber is connected to the α motoneurons and has an inhibiting effect on them. In this way, the contraction of the extrafusal fibers is also inhibited if there is a danger of a muscle tear through overstretching. The Golgi tendon organs represent an important protective mechanism. In addition, the outflow of the α motoneurons reacts with the anterior horn cells (feedback) via an inhibiting intermediary neuron (Renshaw cell). The outflow of the α motoneurons is thus adapted to the pattern of muscle contraction and a certain amount of damping occurs.

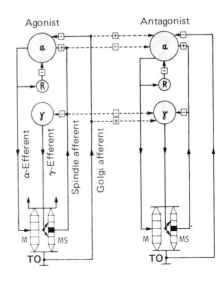

Fig 16
Schematic diagram of proprioceptor reflex arc. M, working musculature; MS, muscle spindle; R, Renshaw cell; TO, tendon organ. Processes of facilitation and inhibition of the α and γ motoneurons of agonists and antagonists (due to stretching of the test muscle) are indicated by the appropriate symbols (+ = facilitation; − = inhibition). From Caspers, H.: Zentralnervensystem, In: Kurzgefasstes Lehrbuch der Physiologie, 4th ed., ed. by Keidel, W. D. (Stuttgart: Thieme, 1975).

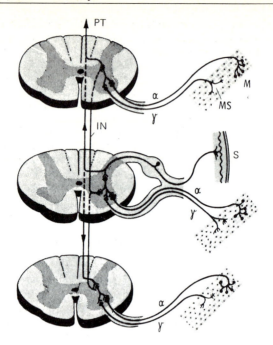

Fig 17   Diagram of exteroceptor reflex arc. PT, ascending and descending posterior tract with reflex collaterals which end at the switching neuron; S, skin with sensory nerve fibers; MS, muscle spindle; M, working muscle with the α motoneuron; IN, intermediary neuron. From Caspers, H.: Zentralnervensystem, In: Kurzgefasstes Lehrbuch der Physiologie, 4th ed., ed. by Keidel, W. D. (Stuttgart: Thieme, 1975).

Because every muscle contracts as a result of spindle stretching when it is stretched, there would be a continual conflict between the agonists and antagonists (e. g., flexors and extensors) if the α and γ motoneurons did not reciprocally influence these muscles. Both motoneurons of the antagonists are inhibited when the agonists are facilitated and vice versa.

1.6.5 Polysynaptic Motor Reflex

In monosynaptic reflexes, the receptors are located in either the muscles themselves or their tendons. Many of the remaining receptors in the body (e. g., touch receptors of the skin, pain and temperature receptors) can also initiate motor reflexes. It is characteristic of this type of reflex, also called exteroceptor reflex, that there is a large variability. This is because reflex time and transmission are highly dependent on the spatial and temporal effects and the intensity of the stimulus. This will become clearer if one thinks about how a person acts when he steps barefoot on a small stone or on a sharp piece of glass. In the case of the stone, he will displace his weight slightly and in such a way that an observer may possibly not even notice. However, with the protective reaction produced by the piece of glass, many muscles are rapidly activated. The scheme of the exteroceptor reflex is shown in Figure 17. Starting from the skin receptors, the α and γ motoneurons are increasingly or decreasingly activated via one or more intermediary neurons, resulting in a coordinated reflex movement.

Within the framework of the fundamentals of motor activity discussed here, the polysynaptic reflex distinguishes itself from the monosynaptic reflex in several ways. First, it demonstrates the phenomenon of summation, i. e., subthreshold stimuli can lead to the initiation of a reflex over a long period of time. A typical reflex of this group is the sneeze reflex. Not every tiny particle which reaches the nasal mucous membrane causes a sneeze. However, the impulses from the receptors are accumulated and initiate the reflex when a definite threshold is surpassed. A further characteristic is the possibility of facilitating and inhibiting the reflex. One can raise the threshold to a certain level, for example, if he has to sneeze while sitting at the piano in a concert. On the other hand, one can also facilitate the reflex by looking into the sun because in addition to the nasal stimulation, there is an extra stimulation of the eyes. A further trait of polysynaptic reflexes is the wider range of motor reactions resulting from the stimulus intensity.

Polysynaptic reflexes are not concerned only with motor activities in which they intervene in the movement pattern as a result of information from the body's exterior. Some of them are associated with special protective functions (cough reflex, eyelid reflex) and some serve certain autonomic functions (evacuation of the bladder and intestine) or reproductive functions (genital reflex).

1.6.6 Pyramidal and Extrapyramidal Systems

After becoming acquainted with the monosynaptie and polysynaptic reflexes, we now look at the higher levels of the motor system. Figure 18 shows the macroscopic division of the CNS. The spinal cord is at the bottom of the drawing; a large part of the α and γ motoneurons which were previously studied are located here. Attached to the spinal cord is the medulla oblongata, the pons and the midbrain, which together form the brain stem. The cerebellum exercises a particularly important motor function, which controls the spatial and temporal coordination of movement. In comparison to the brain stem, man has a highly

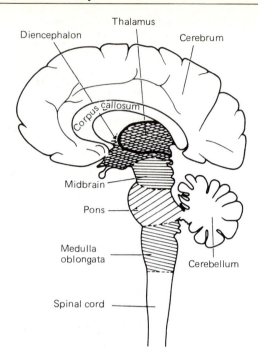

Fig 18   Schematic diagram of the structure of the central nervous system. Medulla oblongata, pons and midbrain together form the brain stem. From Schmid, R. F.: Grundriss der Neurophysiologie (Berlin: Springer, 1974).

developed cerebrum, which is formed from two fairly independent halves. These hemispheres are connected to each other by a bundle of nerve fibers running primarily in the corpus callosum. The diencephalon is that region located between the brain stem and cerebrum. The thalamus, which is particularly important for sensory activities, and the hypothalamus, which is necessary for autonomic control, are both located in the diencephalon.

1.6.6.1 Corticospinal Tract (Pyramidal Tract)

The route of the pyramidal tract is schematically presented in Figure 19. The tracts originate mainly in the precentral gyrus of the motor cortex, i. e., the neurons belonging to them are located there. They pass by the thalamus. In the brain stem, at the so-called pyramid, a number then cross over to the other side and pass down the posterolateral quarter of the spinal cord. A smaller portion runs uncrossed until

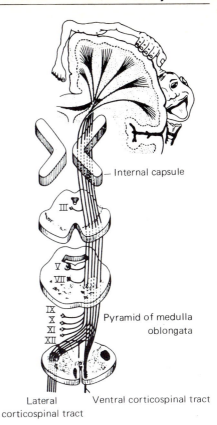

Fig 19
Origin and course of
pyramidal tract in man.
Figure drawn in at top
illustrates size of rep-
resentation fields in precentral gyrus in frontal
section of motor cortex
(motor homunkulus).
From Caspers, A., In:
Kurzgefasstes Lehrbuch
der Physiologie, 4th ed.,
ed. by Keidel, W. D.
(Stuttgart: Thieme 1975).

Internal capsule

Pyramid of medulla
oblongata

Lateral
corticospinal tract

Ventral corticospinal tract

the cervical and thoracic cords. Some of the axons of the pyramidal
tract end either directly at the motoneuron or at intermediary neurons
which can then segmentally affect the motoneurons.

There is a close relationship between cells of the motor cortex and
definite muscles and muscle groups. Each muscle is represented in the
motor cortex by a definite number of specific cells. Those muscles such
as the tongue or hand muscles which have to perform multifaceted
motor tasks have a much larger number of motor cortex cells than
those muscles which primarily carry out holding and supporting functions.

Simplified, it can be stated that the pyramidal tract is responsible for
rapid voluntary movement. It is especially well-adapted for this
because some of its axons go directly to the the α motoneuron, allowing
the information to be transmitted rapidly. Its influence is predomi-

nantly facilitating; this can be inferred from the fact that an interruption leads to a state of flaccid paralysis.

1.6.6.2 Extrapyramidal Tracts

A shot putter who has just contracted the necessary muscles to heave the shot, thus activating his pyramidal system, would completely lose his balance if a second system did not adjust the total compensatory motor activity. This second system is the extrapyramidal system.

Several important extrapyramidal tracts are shown in Figure 20. As was seen with the pyramidal system, many of their cells originate in the

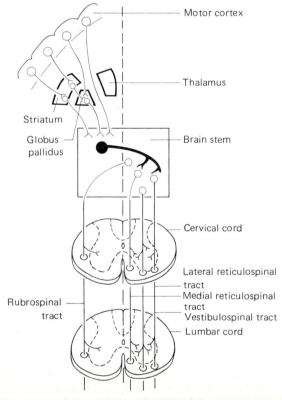

Fig 20   Several important extrapyramidal tracts from supraspinal motor centers in the spinal cord. Most of the extrapyramidal tracts crossover to the contralateral side at the level of the brain stem. From Schmidt, R. F., Ed.: Grundriss der Neurophysiologie (Berlin: Springer, 1974).

motor cortex, but their axons are basically shorter. Along the way, the tracts switch at least once and many switch several times. The synapses are in the basal ganglia of the motor cortex, in the striatum or in the pallidum. Several important related spinal cord tracts are shown in the diagram. In the brain stem, as well as in other switching sites, a series of sensory afferent fibers (e. g., afferents from the vestibular organs of equilibrium) can be connected.

From the morphological arrangement alone, as well as from the many switching sites, it should be clear that we are talking about a more complex and multifaceted system. As we have already seen, synapses are always the sites where the excitation transmission can be inhibited and facilitated and thus modified and controlled.

In summary, the extrapyramidal tracts primarily serve to control holding functions, slow movements and tonus adjustments. In these functions, the predominating influences are those of inhibition; the result of an isolated failure would be spastic paralysis. The extrapyramidal tracts predominantly affect the γ motoneurons; the α motoneurons are primarily activated via the γ loop.

1.6.6.3 Cerebellum

The cerebellum is clearly separated from the rest of the brain and connected to it by thick strands of efferent and afferent tracts. Using axon collaterals and special pathways, the cerebellum receives exact copies of the afferent and efferent information flow going to and from the motor centers. It is in the position to exchange reciprocal bits of information with the cerebrum. It also receives information from all the sense organs. From these connections, it is possible to conclude that the cerebellum has essential coordination functions. It is responsible for the production of a frictionless, goal-oriented execution of the voluntary movements planned by the cerebrum. Furthermore, as a part of the extrapyramidal system, it has to synchronize these voluntary movements with those motor activities associated with muscle tonus, posture and balance.

A complete disconnection of the cerebellum causes a reeling, staggering gait because the muscle movements are not initiated at the right moment and not with the correct level of strength (ataxia). In addition, there is a shaking during goal-oriented, voluntary movement (intention tremor) and an inability to complete rapid movements in succession, such as is necessary when typing (adiadokokinesia).

The pyramidal tract sends axon collaterals to the cerebellum so that it will be informed beforehand about voluntary movements. In this way, it can modify the flow of stimulation in the extrapyramidal system (e. g., to displace the center of gravity), as well as correct the motor

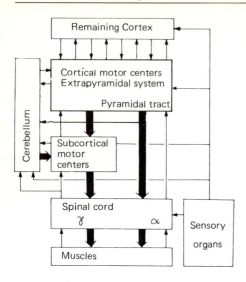

Fig 21
Block diagram of connections between cortical motor centers with other motor centers and with other portions of the cerebellum, muscles and sensory organs. From Schmidt, R. F., Ed.: Grundriss der Neurophysiologie (Berlin: Springer, 1974).

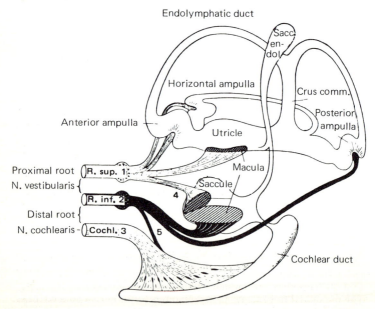

Fig 22   Diagram of the innervation of the labyrinthine sensory organs according to De Burlet. From Schneider, M.: Einführung in die Physiologie des Menschen (Berlin: Springer, 1971).

cortex after receiving sensory feedback. An overview in the form of a block diagram can be seen in Figure 21.

1.6.7 Influence of the Organs of Equilibrium on Motor Activity

In the world of sport, space orientation has an especially important role in all types of acrobatic sports. With these sports, it is as necessary to notice the momentary position of the body in relation to the force of gravity as the linear and rotational accelerations to which the body is subjected.

The macula organs of the inner ear delivers information about the position of the body in relation to gravity. It is composed of the utricle and saccule, in which there is an accumulation of sense organs. Both contain short sensory hairs which project into and are simultaneously held together by a gelatinous mass (Fig 22).

These sensory hairs are supplied by the vestibular nerve. Under normal body positions, the right and left branches of this nerve fire continuous symmetric discharges toward the CNS. However, there is a change in the impulse frequency with changes in the position of the head. When the head is slanted, there is an increase on one side and a decrease on the other. When the head is bent forward or backward, different fibers are stimulated.

The sensory organs signal the position of the head. In animals, they produce a characteristic reflex which always holds the head in a vertical position. This reflex is partially overridden by higher processes in man. The sensory organs have an especially important role during changes in head position relative to the environment, e. g., by keeping a fixation point while the body executes relative movements. Additionally, these sensory organs facilitate the evaluation of rising or falling during progressive acceleration, even when this information is secondarily obtained via changes in muscle tonus.

For sporting activities, it is important that a linear acceleration in the direction of the ground produces maximal tone of the extensors, as can be observed in a rapidly accelerated elevator with "spreading of the toes."

1.6.7.1 The Cupula Organs of the Semicircular Canals

The semicircular canals are situated in the immediate vicinity of the inner ear such that all three spatial planes can be ascertained. They have tiny bony ridges containing a sensory epithelium whose cells are supplied with long cilia clustered together and distributed throughout the endolymphatic canal. During rotary movement, although the semicircular canal moves with the skull, the endolymph of the horizon-

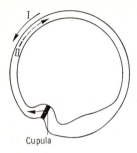

Cupula

Fig 23   Diagram of directional mechanism of the cupula during rotational acceleration. I, direction of rotation of the skull; II, relative movement of semicircular canal fluid. This fluid initially lags behind due to its inertia and then is carried along because of the internal friction. Endolymphatic flow displaces the cupula by this process. From Schneider, M.: Einführung in die Physiologie des Menschen (Berlin: Springer, 1971).

tal semicircular canal lags behind due to its inertia. With continued movement, the endolymph eventually moves at the same speed as the skull. The relative movement of the endolymph transmits the adequate stimulus from the sensory organs of the semicircular canal to the ampulla. This stimulus is formed only by variations in rotary speed of the head (positive or negative rotational acceleration). Figure 23 illustrates the mechanical principle. When the ring is rotated, the endolymph at first lags behind. In this way, the cupula is affected until both velocities are identical. The opposite situation occurs with a sudden stop after rotary movement. After the semicircular canal wall stops rotating, the endolymph continues to move for several seconds; as a result, the cupula is affected on the other side. With each rotary acceleration, there is a transitory movement of the endolymph in the corresponding semicircular canal and an alteration of the excitatory conditions of the cupula; this is then conducted toward the CNS over the vestibular nerve.

The flow of endolymph can also be inappropriately produced when the auditory canal is rapidly cooled. Movement then occurs because of the different specific weight of the variably heated fluid. The cupula signals that a rotary acceleration is occurring, even though one is not present. This is especially dangerous during diving (barotrauma of the ear, p 218) because it can lead to a loss of orientation under water.

The organs of equilibrium have an influence on motor control in that they maintain and restore the normal positions of the body and head. This is done in conjunction with reflexes which receive signals from the retina. The head is thus "justified" relative to the earth's force of gravity (holding and positional reflexes). This reflex also retains a fixation point, so that the environment does not appear to move. The position of the eyes at first lags somewhat behind and then suddenly springs forward into the new position. If this reflex did not exist, the environment would seem to dance back and forth while driving an auto over a bad road.

The organs of equilibrium are also associated with autonomic and emotional centers (limbic system) because an excessive activation produces nausea and defensive reactions; these are termed kinetoses (e. g., sea sickness).

1.6.8 Motor Learning

The acquisition of movement skills is important in almost all aspects of man's life. Although some skills are learned during childhood play, the more complicated or technologic the skills are, the more they require goal-oriented instruction. In the forefront is the idea of "learning by doing." Skills are consciously executed step by step, constantly repeated and then slowly become automatic, i. e., a dynamic stereotype is developed. Because the primary focus of this book is work and sport, let us consider the acquisition of a movement pattern in a ski school. Non-athletes can also remember their learning experiences with driving lessons. Why have we chosen these two examples? Both activities require a large amount of acquired coordination. There is a real difference between the beginner and the skilled performer in that the beginner must consciously execute the movement pattern of each phase, whereas the skilled person just has to give a type of starting signal for the movement pattern and all the rest follows automatically. Even the starting signal can be automatic. For example, if there is an incline in the street or if the number of motor revolutions decreases, the experienced driver shifts down without consciously thinking about it. The skier also adapts his swings automatically to the terrain.

In addition to the classic learning method of actively executing the activity, there are other didactic-methodologic measures which can be used to improve movement skills, including observation, sensory-motor learning and mental practice. Observation is the examination of the correct movement pattern as performed by a teacher or coach, followed by imitation of that pattern. With such modern technical aids as video tape recorders, it is also possible to compares one's own movement pattern with the correct one. As a result, there appears a sort of "actual-desired value" difference, which can be used for correction. Because sensory organs and motor control functions operate together in an external loop, this procedure is also called sensory-motor learning. In this connection, attempts are being made to formulate cybernetic laws (see supplement, p 301) and to make them measurable.

The third form of simplifying the learning of motor skills is mental practice. This is nothing more than imagining a movement pattern that one has already practiced physically or has assimilated through observation and then repeating it in one's imagination without actually executing the movement.

The learning of movement skills with complicated movement patterns requires a great deal of experience because the inexperienced student incorrectly evaluates the feedback from his own proprioreceptors. Every change in direction during skiing requires a displacement of the skier's body weight; on a slippery surface, this requires an active high-low or low-high movement. Although the beginner subjectively means to execute this movement correctly, objectively the situation is different. Furthermore, the abundance of details that must be observed bewilders the student and inhibits the formation of the movement stereotype. Therefore, the experienced teacher will make the student practice the different features of the complicated movement pattern individually. The limbic system also has an effect on the learning process, e. g., when the novice skier is on a steep slope and is afraid, he no longer functions properly.

Motor learning has multidisciplinary aspects. However, this presentation is limited to the basic physiological mechanisms. The observation that a conscious or controlled act can be transformed into an involuntary, unconscious one during the course of the learning process illustrates well that a new reflex mechanism is apparently being formed. The underlying, new combinations are called "conditioned reflexes" and the resulting movement patterns are called "conditioned responses." Conditioned reflexes were first described by Pavlov. He studied autonomic exteroceptor reflexes, whose receptors were located in the mucous membrane of the mouth and which affected the secretion of gastric juices. In order to study this process in isolation, he cut the esophagus while the dog was under anesthesia and brought the proximal and distal ends to the outside at the level of the neck. When the dog ate, the chyme fell to the outside through the proximal stub, while at the distal end there was a gastric tube through which he could withdraw gastric juice.

Pavlov observed that he could create a new reflex if he simultaneously rang a bell at feeding time (he probably used it to call the caretaker of the animals). Basically, Pavlov found that after a certain time period, he needed only the bell signal to initiate the increased secretion of gastric juices. He had therefore reflexly combined two systems, which up to that point did not belong together.

Further research revealed that not only sensory and autonomic systems can be bound together through conditioned reflexes, but that all the other parts of the nervous system can be combined with each other. The difference between the character of the unconditioned, inherited responses based on exteroceptor reflexes and that of conditioned responses is only quantitative and not qualitative. This is especially true for acquired defensive reflexes. Autogenic training is also based on the formation of conditioned responses.

The acquisition of movement skills concerns another form of conditioned reflex which is associated primarily with the efferent portion of the response. A new movement repertoire is formed which can be added to the inherited ones. Those conditioned reflexes which concern only the effector components are also designated as being of this second type because they basically have no parallel with the unconditioned reflexes. If the acquired movement skills operate in a complex manner via a multiplicity of new combinations, then there should be no doubt that we are discussing the formation of conditioned reflexes.

The investigation of the underlying mechanisms is difficult. Most likely, the neurophysiologic representation passes from the motor cortex level (which is also reflected in the consciousness as voluntary movement) over to a subcortical level (perhaps even to the brain stem) and becomes unconscious. Movements which are automatically executed can nevertheless always be called into consciousness. For example, an auto driver who has reacted as a result of a suddenly appearing, dangerous situation, can consciously recall all the details of his response later on.

As a rule, the formation of conditioned responses occurs in several phases. First, there is a unification of the partial actions into a total action. In the second phase, extraneous movements and excessive muscle tension are eliminated. In the third phase, the efferent system is made even more precise. At the beginning, there are always too many responses of extraneous muscles implicated in the work; this can lead to premature fatigue and have a negative influence on coordination. According to recent studies, it appears that it is also important to acquire a certain rhythm to consolidate the dynamic stereotype.

Because in principle each movement skill involves an afferent, a central and an efferent component (in each, time is frequently decisive), it is easy to imagine various types of performance in which the limiting factor can differ greatly from one type to another. In the case of a soccer referee, the external afferent component has the main role, whereas the central component (the decision) and the efferent component (the whistle) are relatively easy to deal with. With chess players, the afferent and efferent components are simple (i. e., the perception of and the moving of the pieces, respectively), but the central component is complicated because it must correctly respond to a whole series of possible moves. There are also many examples of efferent difficulties.

1.6.9 Economical Use of Muscle Strength

The muscles in the body are called on to perform two entirely different tasks. The first is to perform mechanical work via movement against a force. To do this, the muscle must shorten considerably to achieve movement with the help of the bones and joints. The second task is that of holding a force without movement and without performing external work. As an example, when a person stands at rest the

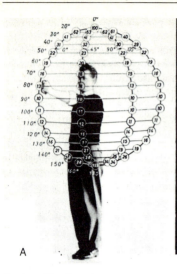

A

Static pulling force in per cent
of body weight; Average value
for right and left arm of
5 subjects.

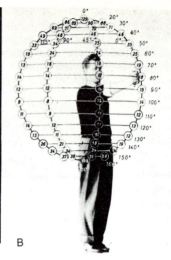

B

Static pushing force in per cent
of body weight; Average value
for right and left arm of
1 subject.

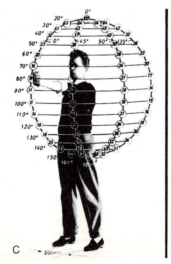

C

Static pulling force in per cent
of body weight; Average value
for right and left arm of 5 subjects

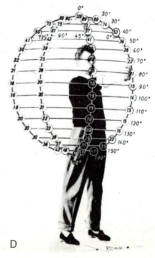

D

Static pushing force in per cent
of body weight; Average value
for right and left arm of 1 subject.

musculature of the legs, buttocks and back has to constantly insure that the bones do not assume another position due to the forces of gravity. Although all muscles are capable of performing both tasks, the muscles of the back are more suited for static work and the arm muscles are more specialized for dynamic work.

The maximal strength which can be mustered by a muscle group against an externally given situation can be ascertained only in an empirical fashion. The force depends essentially on the anatomical realities, especially the lever length between the joint and the point of insertion of the muscle. For practical purposes (e. g., as approximate values for the favorable arrangement of components at work), the maximal pulling and pushing forces expressed in per cent body weight are shown for varying angular positions of the arms (Fig 24). However, one should not believe that the muscle strength of the arms is the limiting factor for maximally applied pushing or pulling force; this is also dependent in the same way on the opposing force which the subject must produce in order not to lose his position.

If large amounts of muscular strength must be exerted, then the opposing force must always be considered in the practical situation. As an example, if a lever is used such that a large amount of force must be applied, then the best angle of application is not 100 degrees but 30.

Most people work in either a sitting or standing position. In both cases, the body must be balanced in an unstable position. All forms of external force disturb this equilibrium. Therefore, the effort must be chosen in the best possible way so that the forces from two equal, counteracting muscle groups will nullify each other (Fig 25). A man can carry a heavy suitcase in one hand much less comfortably than he can carry two suitcases, each weighing one-half as much, in both hands. A favorable arrangement is also present when the force opposing that of a muscle is formed by the force of gravity (Fig 26). In the arrangement described here, the flow of force does not go vertically through the body. In this way, the cyclist involuntarily brings his center of gravity forward when he exerts the muscles of his legs forcefully while sitting on the saddle. The opposing force can also act on a solid point in the environment (Fig 27).

If muscular force is to be applied economically, then the joints should be stressed with the least possible torque. The torque is zero when the tension and the opposing force form an angle of 180 degrees. In this

---

Fig 24   A–D Maximal static force of extended arm in relation to body position; feet are together in A and B, feet are 300 mm apart in C and D. A and C, static pulling force in per cent body weight; average value for the right and left arm of 5 subjects. B and D, static pushing force in per cent body weight; average value for the right and left arm of 1 subject. From Rohmert, W.: Int. Z. angew. Physiol. 18 : 175, 1960.

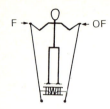

Fig 25   Relation between force (F) and opposing force (OF) during two-armed support (according to E. A. Müller).

Fig 26   During bicycle riding, opposing force is produced by the force of gravity (according to E. A. Müller).

Fig 27   Opposing force needed when moving lever can be supplied by solid wall at the rear (according to E. A. Müller).

situation, the number of muscles which must be tensed is also at its minimum. The active muscle mass increases with the torque of the joints, which the flow of force within the body must overcome. The torque must be counteracted by a muscular torque of the same magnitude if the equilibrium is to continue.

The force-flow relationships are especially important in athletic exercises, e. g., during gymnastics. The analysis of this is the task today of a separate discipline, biomechanics. The reader is referred to the special literature of this science.

1.7 Energy-Supplying Processes for Muscle Activity

1.7.1 Role of the High Energy-Phosphates

During the discussion of electromechanical coupling, it was seen that a muscle contraction occurs only when there is a drop in ATP, which is thus the transmitter of energy to the contractile proteins. Adenosine triphosphate is a mononucleotide made of the purine base adenine, ribose and three linear phosphates linked in series, all of which are linked together through two acid anhydride bonds. Related to these bonds are ADP with two phosphates and adenosine monophosphate (AMP) with one phosphate.

The biologic importance of ATP lies in the two acid anhydride bonds which are easily hydrolyzed, thus releasing considerable amounts of energy. The only form is chemical energy, which can be converted into mechanical work by the contractile proteins; this energy is liberated when the terminal phosphate group is hydrolytically split off. Resting skeletal musculature contains about 2.4 mmol/100 gm dry weight ATP. Another energy-rich phosphate found in muscle is creatine phosphate (CP). In resting skeletal musculature of man, there is a concentration of 6.8 mmol/100 gm dry weight. Creatine phosphate cannot be used directly by muscle for energy but only indirectly with the help of the Lohmann reaction:

$$CP + ADP \rightleftarrows Cr + ATP. \tag{2}$$

This reaction is catalyzed by the enzyme creatine phosphokinase (CPK), which is so active that the ADP present during the contraction is synthesized into ATP during the contraction phase. For this reason, the concentration of ATP remains relatively constant until the levels of CP are completely exhausted. Moreover, the enzyme activity is essentially adapted to the maximal velocity of ATP utilization by the contractile proteins. Its activity is about 10 times higher in skeletal muscle than in heart muscle.

Without refilling the CP reserves, the energy delivered by ATP is sufficient for only 2–3 contractions. From the CP stores themselves, a work load of 89 kpm/kg muscle can be undertaken without replenishment from aerobic and anaerobic metabolism. During the high-level performance by skeletal muscle of 10 kpm/second · kg muscle (corresponding to a 100–meter run in 10 seconds), the CP stores are adequate for about 9 seconds.

1.7.2 Degradation of Carbohydrates

When muscle has used its stock of ATP or CP and has lost its energy to an external source, it now has two possibilities to refill its stock of energy-rich phosphates: aerobically through the biologic oxidation of nutrients or via the anaerobic degradation of glycogen to lactic acid (glycolysis). The energy sources of muscle are schematically demonstrated in Figure 28 with the use of 1 gm frog muscle.

Carbohydrates are stored as glycogen in the sarcoplasm and in the liver cells. Another portion is dissolved in the blood as glucose. In venous blood, there is about 60–80 mg/100 ml; it is about 20% higher in arterial blood.

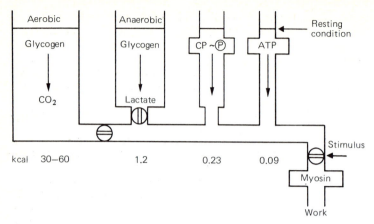

Fig 28   Schematic representation of energy reserves of 1 gm frog muscle. From Rappoport, S.: Medizinische Biochemie, 3rd ed. (Berlin: VEB Volk und Gesundheit, 1965).

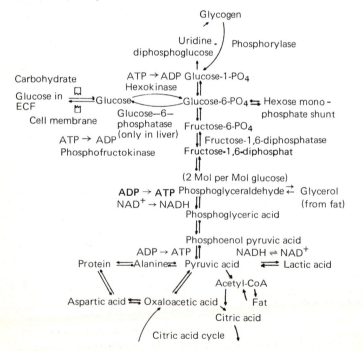

Fig 29   Schematic of carbohydrate and fat degradation.

The basic principle of energy production in the organism is oxidation of foodstuffs with oxygen; energy can also be temporarily produced by chemical reconstruction of the molecular structure. Oxidation is defined as the combination of a substance with oxygen or the release of hydrogen or electrons. Biologic oxidations are catalyzed by enzymes in such a way that a certain enzyme is responsible for a certain reaction.

An overview of carbohydrate metabolism in the cell and the most important enzymes can be seen in Figure 29. The details exceed the realm of this book but can be found in textbooks of biochemistry. The raw materials are glycogen (above) and the glucose in the extracellular fluid (ECF). Glycogen is found within the cell, whereas glucose can pass through the membrane.

Resting muscle cells contain a considerable reservoir of glycogen. The question can then be raised as to what actually controls the degradation. From the picture, it can be seen that the degradation pathway always causes an accumulation of phosphorus, i. e., degradation can occur only when phosphorus is available from the reduction of ATP. As was seen on p 15, a big drop in ATP occurs only when the electromechanical coupling is activated by the stimulation of the muscle membrane. Thus, this is an example of a self-regulating feedback system.

On the way to pyruvic acid (pyruvate), a small amount of energy is gained via the synthesis of ATP. At this point, the degradation pathway branches out. It is now possible to use either the "anaerobic" pathway, producing lactic acid, or biologic oxidation will occur "aerobically" over the citric acid cycle and the respiratory chain.

1.7.3 Citric Acid Cycle and Respiratory Chain

Pyruvic acid ($C_3H_4O_3$) is formed from glucose ($C_6H_{12}O_6$) in the following manner:

$$C_6H_{12}O_6 \rightleftarrows 2C_3H_4O_3 + 2\ NADH + 2H^+. \tag{3}$$

The resulting hydrogen is bound to the coenzyme (active group of an enzyme) NAD (nicotinamide adenine dinucleotide). The $NADH + H^+$ passes on its hydrogen to the enzymes of the respiratory chain in the mitochondria, where it is combusted in the presence of oxygen (O_2) to water (H_2O). A portion of the energy liberated is stored as ATP. If O_2 is not present or if the activity of the enzymes in the citric acid cycle or respiratory chain is relatively insufficient for a high production of H_2, then the H_2 from the $NAD \cdot H_2$ is transferred to the $C_3H_4O_3$ and the result is lactic acid:

$$C_3H_4O_3 + NADH + H^+ = C_3H_6O_3 + NAD. \tag{4}$$

The resulting $C_3H_6O_3$ dissociates into H^+ ions and lactate ions, which are channeled into the extracellular space and subsequently into the blood. The hydrogen ions are partially buffered (see p 151) and the

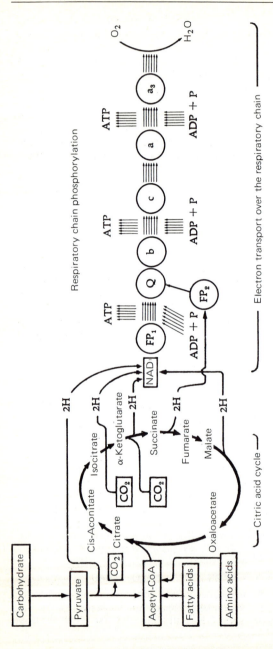

Fig 30   Schematic review of oxidation of carbohydrates, fatty acids and amino acids. Symbols FP_1 and FP_2 represent the NADH and succinate dehyrogenases, Q is coenzyme, and Q plus b, c, a and a_3 symbolize the cytochromes. From Lehninger, A. L.: Bioenergetik, 2nd ed., (Stuttgart: Thieme, 1974).

lactate is then either partially combusted by the liver, heart or resting skeletal muscle or resynthesized into glucose. The $C_3H_4O_3$ also dissociates into H^+ and pyruvate. However, only a small amount of pyruvate leaves the cell as most diffuses into the mitochondria, where it is immediately coupled with coenzyme A (CoA). Carbon dioxide is split off and the H^+ is loaded onto NAD, while the rest remains as acetyl-CoA (activated acetic acid). Subsequently, $NAD+H^+$ yields its H_2 to the respiratory chain for combustion. Acetyl-CoA is converted and degraded in several steps in the citric acid cycle, resulting in the accumulation of H_2O (Fig 30). The most important result for the energy-yielding metabolism is the formation of H_2 and the end-product CO_2.

$$CH_3\text{-CO SCoA} + 3H_2O = 2CO_2 \qquad + 8\,H \qquad + \text{HS-CoA}$$
$$\text{(Acetyl-CoA} + 3\,\text{water} = 2\,\text{carbon dioxide} + 8\,\text{Hydrogen} + \text{CoA).} \qquad (5)$$

The CO_2 is carried away and the H_2 is taken up by the coenzymes and passed on to the respiratory chain enzymes.
The oxidation of H_2 in the respiratory chain with $\frac{1}{2}O_2$ to H_2O is the decisive reaction of the energy-producing metabolism. The respiratory chain represents a multienzyme system which is built into the walls and cristae of the mitochondira in a closely arranged manner and which terminates with the cytochromes (a, b, c). With the help of these enzyme chains, electrons (–) are removed from the occurring H_2. In this way, they become positively loaded H^+ ions. Oxygen is charged with the electrons and becomes a negatively loaded O^{--} oxygen ion. Both ions are then bound together as water:

$$2H^+ + O^{--} = H_2O + \text{Energy.} \qquad (6)$$

Approximately 40% of the resulting energy is chemically stored as ATP. Thus, biologic oxidation is an electron displacement between H and O. The stoichiometric and energetic balance of the complete degradation of 1 mol free glucose can be described as follows:

Without O_2: $C_6H_{12}O_6$	$= 2C_3H_6O_3$	$+ \; 36$ kcal	(2 ATP yield) (7)

With O_2: $C_6H_{12}O_6$	$= 2C_3H_4O_3 + 4H +$ 36 kcal	(2 ATP yield)	(8)
$2C_3H_4O_3 + 6H_2O$	$= 6CO_2 + 20\,H \; + \; 50$ kcal	(2 ATP yield)	(9)
$12H_2 + 6O_2$	$= 12H_2O \qquad + 588$ kcal	(34 ATP yield)	(10)

$C_6H_{12}O_6 + 6O_2$	$= 6CO_2 + 6H_2O + 674$ kcal	(38 ATP yield)	(11)

When 1 mol glucose in its stored, energy-rich form (glycogen) is degraded and converted, 39 mol ATP are produced. The O_2 required for equation 10 comes from the air. The diagram makes it clear that the greatest yield of energy and ATP comes from the oxidation of H_2 to

H_2O. The ATP is formed from ADP and the phosphoric acid residual (P_i), with the assistance of the oxidation energy; this process is called oxidative phosphorylation.

The amount of available ADP and P_i is a control mechanism of the respiratory chain. The mitochondria terminate their oxidative energy production when the oxygen pressure goes below a value of 5 mm Hg. An O_2 deprivation or an insufficient level of enzymes in the citric acid cycle or respiratory chain will result in severe reductions in energy production. The citric acid cycle will function normally only when the respiratory chain operates intact.

The degradation of fatty acids also occurs at the level of acetyl CoA in the citric acid cycle; further degradation and ATP yield are thus identical. Fatty acids contain fewer molecules of O_2 than glucose. As a result, more O_2 must be delivered via the blood during their combustion; in general, trained persons can do this better. Fatty acids cannot be converted anaerobically.

At rest, skeletal muscle combusts practically only carbohydrate. During heavy work, muscle also uses free fatty acids. Thus, their combustion covers a considerable portion of the energy requirement. The reader should refer to the literature on physiologic chemistry for more details.

1.7.4 Interchange between the Energy-Supplying Systems

The energy-rich phosphates, primarily ATP, assume a central role in the metabolic events of resting and working muscle. The degradation of ATP, which is initiated by an excitation and the resulting liberation of calcium ions from the lateral sacs, gives muscle the ability to react in fractions of a second. Thus, ATP represents an energy reserve which is capable of being mobilized in a short period of time. All of the other reactions described so far serve to replenish this energy reservoir; simultaneously, they control their own refilling with a sort of control process. The supply of ATP alone is sufficient for only a few contractions. The next reservoir is CP, which is capable of rapidly replenishing the unloaded ATP stores.

This refilling begins even during the contraction. As we have already seen, cell respiration (or oxidative metabolism) begins simultaneously and continues only as long as inorganic phosphate and ADP are present. When oxidative metabolism cannot continue due to a lack of oxygen as an H^+acceptor or overcharging the capacity of several enzymes, then the process of glycolysis is brought in to resynthesize ATP and CP.

The dynamics of the process are illustrated in Figure 31. The isolated gastrocnemius muscle of a dog was supplied with oxygenated blood and the time course of oxygen intake was measured from the beginning

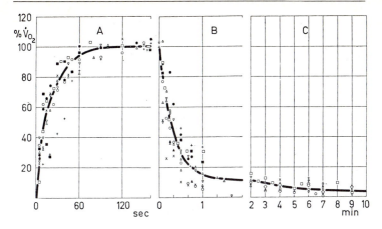

Fig 31   A, time course between rapid increase in work of muscle group and increase in oxygen intake required (100%) for aerobic work. B, time course of reduction in oxygen intake of isolated muscle group after exercise. C, continuation of B, with compressed time scale on the abscissa. From Di Prampero P. E. and Margaria, R.: Pflügers Arch. 304 : 11. 1968.

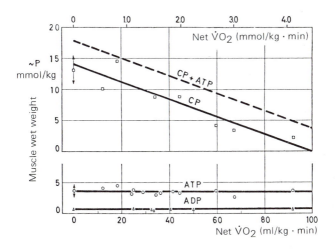

Fig 32   Reduction in energy-rich phosphates (creatine phosphate + ATP) in muscle as function of net oxygen intake. Lower part shows that ATP and ADP concentrations are essentially independent of oxygen intake. From Di Prampero P. E. and Margaria, R.: Pflügers Arch. 304 : 11, 1968.

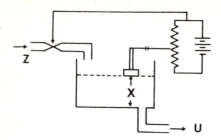

Fig 33
Functional model for explaining oxygen debt. More detailed explanations can be found in the text.

of work, through the steady-state exercise period and until the return to resting values after exercise (Fig 31, A). The oxygen content in the blood entering and leaving the isolated muscle was also determined. On the abscissa is the time in seconds; the time point O represents the beginning of a constant work load induced by stimulation of the motor nerve. In Figure 31, B the time period O represents the end of work; the figure continues in Figure 31, C with a compressed time axis. On the ordinate is O_2 intake in per cent of the maximal value obtained during aerobically performed work. Oxygen intake first attains its steady-state value after a certain period of time. This means that the muscle must have entered into an oxygen deficit because the energy for the performance of work had to be immediately available. In reality, a reduction in energy-rich phosphates can also be seen during this aerobic work; this is inversely proportional to the steady-state O_2 intake (Fig 32).

The reason for this so-called alactacid oxygen debt is the slow rise in aerobic metabolism. This pattern can be clarified with the help of a functional model (Fig 33). There is a fluid volume (X) in a water tank and a continual flow of water into the basin (Z) and an outflow (U). If there is a drop in water level caused by sudden increase in U, then Z will be adjusted by a float and a regulator until there is once more a balance between U and Z. In this model, X symbolizes the concentration of energy-rich phosphates.

When there is a drop in concentration, then there will be an accumulation of inorganic phosphate, which in turn causes an increased resynthesis of energy-rich phosphates. If we assume in an equation that a change in the aerobic metabolic variable (Z) should always be inversely proportional to a change in the concentration of energy-rich phosphates (X), then we obtain (Z_O = initial value; $Z_{(t)}$ = time function):

$$dZ = -adX. \tag{12}$$

Through integration of this equation, we obtain:

$$Z = -aX + Z_O. \tag{13}$$

The change in volume in the tank depends on the inflow rate (Z) and the outflow (U):

$$\frac{dX}{d_t} = Z - U. \tag{14}$$

If equation 13 is inserted into equation 14, then:

$$\frac{dX}{d_t} = -U + Z_O - aX. \tag{15}$$

In the case of a rapid rise in energy requirement (U), the solution of the differential equation reads:

$$X = Ua^{-1}e^{-at} + (Z_O - U)a^{-1}. \tag{16}$$

If equation 16 is solved for Z with the application of equation 13, then:

$$Z_{(t)} = U(1-e^{-at}). \tag{17}$$

In the steady-state ($\frac{dX}{d_t} = 0$), inflow equals outflow, i. e., there is a state of equilibrium between degradation and synthesis. This equation means that the energy turnover adapts exponentially to the energy requirement. The model thus fulfills theoretical and practical conditions. The exponent a in Figure 31 has a value of 0.0345. The reciprocal value is called the time constant and has a value of 29 seconds. The corresponding time constants of oxygen intake patterns measured in man at the onset of work are somewhat higher, but are still in the same range.

From equations 12 and 13, it follows that the increase in O_2 debt, which corresponds to a reduction in energy-rich phosphates (X), must be proportional to the metabolism (U) when a steady state is reached. Again, the proportionality factor a = 0.0345/sec. For each increase in oxygen intake of 100 ml/kg · min, there must be a corresponding increase in oxygen debt:

$$dX = \frac{100}{0.0345 \cdot 60} = 48 \text{ ml } O_2/kg. \tag{18}$$

In addition, one can examine whether the reduction in phosphate concentration measured in the steady state (Fig 32) agrees with the value of the oxygen debt. On page 49, we determined that 38 mol ATP are produced when 1 mol glucose is combusted with 6 mol O_2; this means that 38 mol ATP correspond to a volume of 6 × 22.4 l = 134.4 l O_2. According to this, 1 mol O_2 debt corresponds to a reduction in phosphate content of 0.28 mmol. The reduction of 13 mmol/kg muscle agrees well with the O_2 debt measured and calculated from the increased oxygen intake of 100 ml/kg/min.

The relationship becomes more complicated when the anaerobic energy production is considered. Figure 34 shows the pattern of

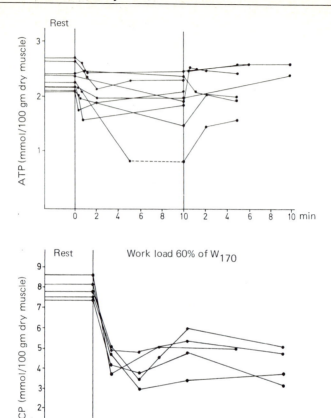

Fig 34   Effect of exercise on breakdown and resynthesis of ATP (above) and on concentration of CP (below). Work load was 60% of work load at which each subject attained a heart rate of 170 beats per minute (W_{170}). From Bergström, J.: Circ. Res. 20, suppl. 1 : 91, 1967.

decrease in CP stores of subjects working on the bicycle ergometer at a point just above their threshold for prolonged work ("Dauerleistungsgrenze"*, p 242); at this work load, a part of the energy needed is

---

* Translator's note: The term "Dauerleistungsgrenze" cannot be directly translated into any term in English which is common to exercise physiology. Nevertheless, it is essentially the same as the "anaerobic threshold", as defined by Wasserman, K. et al.: J. Appl. Physiol. 35, 236, 1973.

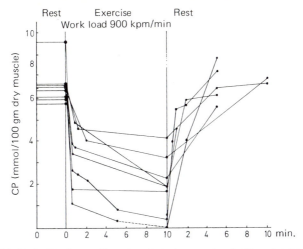

Fig 35 Reduction in creatine phosphate in muscle of different subjects at work load of 900 kpm/minute. From Bergström, J.: Circ. Res. 20, suppl. 1 : 91, 1967.

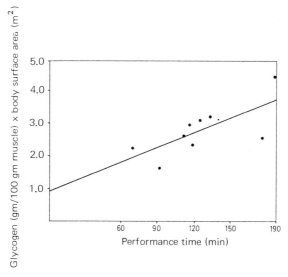

Fig 36 Relationship between muscle glycogen concentration before work and performance time at work load which is 60% of load producing a heart rate of 170 beats per minute. Data from 9 subjects. From Hultman, E.: Circ. Res. 20, suppl. 1 : 99, 1967.

supplied via the anaerobic pathway. Samples of muscle tissue were taken with a needle biopsy from the thigh. The CP was reduced at the onset of work and then remained constant. There was a small drop in the simultaneously measured ATP stores.

Despite the anaerobic energy production, phosphate reserves were held constant over the entire working period.

If the work load is still higher (15 kpm/second), then the reservoir level of CP decreases with time (Fig 35). At this range, it seems that the total aerobic and anaerobic energy production is no longer sufficient to completely replenish phosphate stores during the relaxation phases. Here also, ATP stores are generally constant; they break down only with exhaustion of the subject.

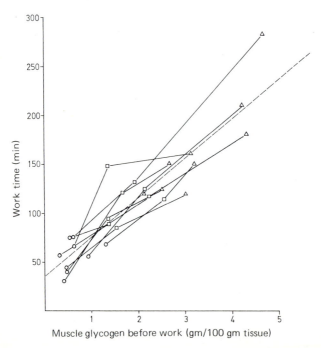

Fig 37   Relationship between muscle glycogen concentration before work and endurance of 9 subjects. Each subject was investigated 3 times with different concentrations of muscle glycogen at the start of work. Squares represent studies performed after a mixed diet, circles after a carbohydrate-free diet and triangles after a carbohydrate-rich diet. Dashed line is the regression line. From Hultman, E.: Circ. Res. 20, suppl. 1 : 99, 1967.

From Figure 36 as well from Figure 37, it is clear that levels of muscle glycogen can limit maximal working time. In this case, all subjects worked at 75% of their maximal aerobic capacity but had different initial values of muscle glycogen. It is obvious that there is a linear relation between muscle glycogen content and maximal possible working time and that this relationship is basically independent of the diet eaten before the work test.

1.7.5 Restitution of Reserves after Exercise

As long as CP is available, ATP stores will be replenished during each contraction. The restitution of CP reserves after aerobic work occurs with the same time function as it decreases; this can be clearly seen in Figure 31. The equations which were used for the functional model are also valid here. When the preceding work was partially anaerobic, then most of the CP is replenished immediately after exercise. However, this restitution is not complete because the local metabolism remains elevated in order to resynthesize muscle glycogen; this synthesis occurs relatively slowly. Figure 38 illustrates the restoration of glycogen reserves after eating a carbohydrate diet. The solid line shows the glycogen content of the musculature of one leg (after performing one-legged bicycle ergometer work for one day) during a time interval of 3 days after the exercise. After one day, the glycogen content has returned to its initial value. The surprising thing is that it then continues to rise well above its normal level. The normal level can be seen from the curve of the other, non-working leg. The restoration of

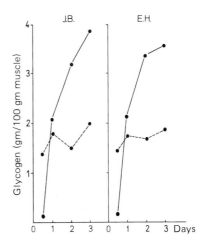

Fig 38
Effect of carbohydrate diet on muscle glycogen concentration after exhausting work. Work consisted of one-legged pedalling on bicycle ergometer until exhaustion. Glycogen content in musculature of working leg is represented by the solid line; that of nonexercised leg is shown by the dashed line. Three days after exhausting work, glycogen content in working leg is about double that of nonworking leg. From Hultman, E.: Circ. Res. 20, suppl. 1 : 99, 1967.

glycogen is influenced markedly by the type of nutrition. From Figure 39, it can be seen that over the 3-day period of the experiment, the glycogen content was not restored when the subjects fasted or ate a carbohydrate-free diet. When they were again given carbohydrate after the third day, the glycogen content increased rapidly.

As we will see in a later section (p 134), the breakdown of energy reserves is closely related to heart rate and the adjustment of blood flow. We will therefore return to this with a discussion of the regulation of circulatory variables.

1.7.6 Muscle Fatigue as a Disturbance of Biochemical Equilibrium

Muscle fatigue is defined as a condition in which the work capacity of muscle is decreased; recovery is the condition where it increases once more. The degree of fatigue can be characterized by the quantitative extent of fatigue and the degree of recovery by the restitution of the resting condition. With each muscle contraction, the biochemical equilibrium is disturbed. If only aerobic work is performed, a reduction in the energy reserves of ATP would occur for a tiny moment during the contraction; according to the definition above, this could also be labeled fatigue. This fatigue, however, is completely compensated during recovery (replenishment of ATP stores) of the muscle, i.e., there is no residual fatigue. Fatigue and recovery counterbalance

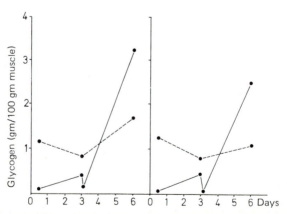

Fig 39   Influence of nutrition on resynthesis of muscle glycogen. Symbols are same as in Figure 38. Subjects ate only fat and protein for first 3 days; thereafter, they ate carbohydrates. As long as only fat and protein were provided, muscle glycogen was not resynthesized. From Hultman, E.: Circ. Res. 20, suppl. 1 : 99. 1967.

each other. On the other hand, if the work is carried out either completely or partially anaerobically, then fatigue cannot be completely compensated in the recovery phase. Thus, there is some residual fatigue, which is removed only after exercise during restitution of the resting condition. Muscle fatigue and muscle recovery are expressions of a disturbed metabolic equilibrium in the musculature. The greater the extent of energetic imbalance, the greater will be the muscle fatigue. For more information about peripheral fatigue, see page 249.

1.8 Fundamentals of Energy Metabolism Measurement

When considered over a sufficiently long period of time, biologically utilizable energy comes from the biologic oxidation of foodstuffs. Anaerobic energy production acts like a reservoir which can briefly supply the necessary energy, but which must then be restocked.

It is thus theoretically and practically possible to measure energy metabolism in several ways. Either the amount of heat given off by the body can be determined (direct calorimetry) or this can be calculated in an indirect way from the oxygen intake and carbon dioxide production (indirect calorimetry). Both procedures, when properly applied, show similar results.

1.8.1 Direct Calorimetry

Direct calorimetry largely depends on control technology because the temperature (among other things) of the measuring apparatus must be held exactly constant. Therefore, it is not applicable for the practical purposes of work and sport physiology. Its principle is based on the concept that the total heat given off by conduction, radiation and evaporation can be measured. Earlier installations used a chamber whose walls of double thickness had water flowing through them. The heat production was determined from the amount of water per unit of time which flowed through the walls and the temperature difference between water coming in and going out. There is a special difficulty in this arrangement, in that the respiratory gases in the enclosed chamber had to be continually renewed; this led to slight errors in temperature. In addition, such an installation is static. Newer direct calorimeters use a series of thermocouples to measure the amount of heat given off. Because of the intricate manipulations required, direct calorimeters are used only for the solution of special scientific problems. Historically, their importance lay in the demonstration that the law of conservation of energy was also valid for biologic systems.

1.8.2 Indirect Calorimetry

Indirect calorimetry is based on the concept of a stoichiometric relationship between the combusted foodstuff, the oxygen required for that process and the carbon dioxide produced. Although the basic stoichiometric equation for carbohydrate combustion was already given in equation 11 on page 49, it is repeated here:

$$C_6H_{12}O_6 + 6\,O_2 = 6\,CO_2 + 6\,H_2O + 674\ kcal.$$

Quantitatively, 180 gm glucose (180 is the molecular weight of glucose) combusts with 6 mol O_2 (1 mol O_2 = 22.4 l; 6 × 22.4 = 134.4 l O_2) to produce 6 mol H_2O and 134.4 l CO_2. The caloric equivalent of 1 l O_2 is obtained from the calories which become available from the combustion of the oxygen consumed.

The caloric equivalent of oxygen for glucose combustion (E_{Gl}) in kcal/l O_2 is thus:

$$E_{Gl} = \frac{674\ kcal}{134.4\ l\ O_2} = 5.01\ kcal/l\ O_2. \tag{19}$$

A similar calculation can be made for fat, e. g., for tripalmitin. The basic formula of tripalmitin is $C_{51}H_{98}O_6$. The combustion process is formulated as follows:

$$2\ C_{51}H_{98}O_6 + 145\ O_2 = 102\ CO_2 + 98\ H_2O + 15.250\ kcal. \tag{20}$$

Using the same calculations as above, the caloric equivalent of oxygen for combustion of tripalmitin (E_F) in kcal/l O_2 is:

$$E_F = \frac{15.250\ kcal}{145 \times 22.4\ l\ O_2} = 4.69\ kcal/l\ O_2. \tag{21}$$

The same calculations can be done for protein. However, this is more complicated because the end product of protein metabolism is not just $CO_2 + H_2O$, but also urea, which still contains energy.

From the simple gross calculations, it is clear that the caloric equivalent of oxygen is not a constant variable, but that it depends on which foodstuff is being combusted. If to simplify matters, protein metabolism is disregarded, it is obvious that the equivalent varies from 4.69 to 5.01 kcal/l O_2. To find which foodstuff is being used, apply the knowledge that a molecule of fat contains less O_2 than a molecule of carbohydrate. As can be seen in equation 11, the volume of O_2 required for glucose combustion is the same as the volume of CO_2 produced. With tripalmitin combustion, the ratio in equation 20 is 102 : 145 = 0.703.

The ratio between the volume of CO_2 produced and the volume of O_2 consumed in the same time period is called the respiratory exchange ratio:

$$R = \frac{CO_2}{O_2}. \tag{22}$$

Table 1   Foodstuff Combustion and Heat Production, from Lang-Ranke: Stoff-wechsel und Ernährung (Berlin: Springer, 1950)

1 gm	O_2 Consumption (ml)	CO_2 Formation (ml)	Respiratory Quotient	Heat Production (Kcal)	Heat Value of 1 l O_2	Kcal/l of 1 l CO_2
Carbo-hydrate	828.8	828.8	1.000	4.183	5.047	5.047
Fat	2019.3	1427.3	0.707	9.461	4.686	6.629
Protein	962.3	773.9	0.804	4.316	4.485	5.579

By using R, it is possible to estimate which foodstuffs are being combusted in the body, as well as which caloric equivalent should be used with the oxygen requirement in order to find out the correct energy turnover.

These considerations cannot be applied exactly to the method of indirect calorimetry because we do not eat only glucose and tripalmitin, but a combination of carbohydrate, fat and protein. These foodstuffs themselves further represent mixtures, whose composition will differ depending on whether they come from animal or plant sources. To that extent, indirect calorimetry has both empirical and computational elements.

The empirical values are determined from a standard average diet. The error that can result from this method (that one might alter one's eating habits), is certainly less significant than other errors, which will be discussed later. Table 1 gives a summary of how the variables discussed here change during the biologic combustion of foodstuffs.

Table 2   Relationship between Respiratory Quotient (R) and Caloric Equivalent (Cal.Eq.)

R.	Cal.Eq.	R.	Cal.Eq.	R.	Cal.Eq.
0.70	4.678	0.80	4.801	0.90	4.924
0.71	4.690	0.81	4.813	0.91	4.936
0.72	4.702	0.82	4.825	0.92	4.948
0.73	4.714	0.83	4.838	0.93	4.960
0.74	4.727	0.84	4.850	0.94	4.973
0.75	4.739	0.85	4.863	0.95	4.985
0.76	4.752	0.86	4.875	0.96	4.997
0.77	4.764	0.87	4.887	0.97	5.010
0.78	4.776	0.88	4.900	0.98	5.022
0.79	4.789	0.89	4.912	0.99	5.034
0.80	4.801	0.90	4.924	1.00	5.047

If the diet were composed entirely of fat and carbohydrate, then it would be simple to say on the basis of R, whether only fat (R = 0.7), only carbohydrate (R = 1.0) or 50% fat and 50% carbohydrate (R = 0.85) was combusted. In reality, this calculation would only be disturbed by that portion of calories associated with the combustion of protein. With degradation of protein, not only CO_2 and H_2O appear, but also urea, a compound known to contain nitrogen. Because on the average protein contains 16% nitrogen, the amount of nitrogen found in the urine is multiplied by 6.25 to find out the protein metabolism.

Under normal dietary conditions, protein metabolism accounts for 10–15% of the basal metabolism. Because the exercise metabolism is primarily covered by carbohydrate and secondarily by fat, the fraction of protein in the total metabolism during exercise can be disregarded. Therefore, for practical purposes, it is not necessary to determine the amount of urinary nitrogen. In the practice of work and sport physiology, protein metabolism is ignored completely. Table 2 shows the values of the caloric equivalents associated with each R under these conditions. By eliminating protein metabolism, the measurement of basal metabolism is always about 1–1.5% too high; the working metabolism, however, is essentially correct.

1.8.3 Important Sources of Error in Indirect Calorimetry

Two important sources of error can upset the measurement of energy metabolism so that under certain conditions, the caloric equivalent of the measured R should not be estimated. One of these error sources is metabolic, the other respiratory. We derived the caloric equivalent from the combustion of both fat and carbohydrate, the oxygen intake required and the carbon dioxide produced. What was not taken into consideration was the fact that both of these foodstuffs also can be converted into one another, resulting in considerable shifts in R. If carbohydrate is converted into fat, then R can assume values which are higher than 1.0; in the opposite case (e. g., during starvation), it can show relatively low values.

More frequent than the "metabolic" error is the "respiratory" one that is made during metabolic measurements. The preceding discussions were always based on cellular metabolism. Because for practical reasons, the oxygen required and the CO_2 produced are measured only via the external respiration and not via determination of the arteriovenous O_2 and CO_2 differences present on entering and leaving the tissue, it is tacitly assumed that neither O_2 nor CO_2 was added or taken away on the way from the cell to the respiratory tract. In the case of carbon dioxide, which can be stored in the body in relatively large amounts, this condition is present only when the arterial CO_2 pressure remains exactly constant. Hyperventilation can displace R until a new steady-state between production and loss is reached. This often is found if an inexperienced subject using a mouthpiece is connected to a respiratory apparatus. Even more impressive are the time-dependent shifts in R caused by increasing hyperventilation during exhausting muscular exercise. Finally, fixed acids, such as those found during anaerobic exercise, can force bicarbonate out of its compounds in the blood and tissue, resulting in a falsified R. Thus, it is not possible to estimate metabolism from actual R values; this estimation can be done only when R is averaged over a long period.

Methods of measuring energy metabolism are treated in the supplement (p 309).

1.9 Energy Metabolism and Physical Work

From a work physiology standpoint, the energy metabolism of the human body should always be measured when the energetic demands or the economics of a type of work are to be evaluated. The total energy metabolism during work is composed of several individual types of metabolism. Firstly, there is a permanent basic energy metabolism which is independent of whether an individual works or not. This is clinically called the basal metabolism. Secondly, a part of the energy metabolism is needed for digesting and converting foodstuffs. This fraction includes the energy needed to maintain the body. Finally, there is the energetic cost required for the work itself.

1.9.1 Basal Metabolism

Basal metabolism is that metabolic quantity measured under strict conditions; these include fasting for 12 hours with a previous protein restriction for at least 2 days, complete physical rest and a neutral temperature. For healthy individuals, the values are basically regulated by sex, age and body weight and there is little interindividual variation. The normal levels of metabolism can be read from Table 3.

1.9.2 Resting Metabolism

Resting metabolism represents that metabolism measured at complete body rest without adhering to the special basal metabolic conditions. Work physiologists measure resting metabolism before work. Depending on the external conditions, it is usually 10–15% above the values for basal metabolism.

1.9.3 Work Metabolism

Work metabolism is classified as the difference between the total metabolism during work and resting metabolism. It effectively represents the amount of energy needed for the work performed because the resting requirements have been eliminated.

1.9.3.1 Pattern of Oxygen Intake during Physical Work

The course of oxygen intake varies, depending on whether a small or a large amount of work is performed. For predominantly aerobic work,

Table 3   Harris-Benedict Standards for Determination of Normal Values of Basal Metabolism. Kcal/h can be obtained by adding value from Chart 1 to corresponding value in Chart 2 (Harris and Benedict in: Documenta Geigy, Wissenschaftliche Tabellen, 6th Ed. [Basel: Geigy, 1962])

Chart 1

Weight in Kg	Total Calories/Hour Men	Women	Weight in Kg	Total Calories/Hour Men	Women
12	9.7	–	72	44.0	56.0
14	10.8	–	74	45.2	56.8
16	12.0	–	76	46.3	57.6
18	13.1	–	78	47.5	58.4
20	14.3	–	80	48.6	59.2
22	15.4	–	82	49.7	60.0
24	16.6	–	84	50.9	60.8
26	17.7	37.6	86	52.0	61.6
28	18.8	38.4	88	53.2	62.4
30	19.9	39.2	90	54.3	63.2
32	21.1	40.0	92	55.5	64.0
34	22.2	40.8	94	56.6	64.8
36	23.4	41.6	96	57.8	65.6
38	24.5	42.4	98	58.9	66.4
40	25.7	43.2	100	60.1	67.2
42	26.8	44.0	102	61.2	68.0
44	28.0	44.8	104	62.4	68.8
46	29.1	45.6	106	63.5	69.6
48	30.3	46.4	108	64.7	70.4
50	31.4	47.2	110	65.8	71.2
52	32.6	48.0	112	67.0	72.0
54	33.7	48.8	114	68.1	72.8
56	34.9	49.6	116	69.3	73.6
58	36.0	50.4	118	70.4	74.4
60	37.2	51.2	120	71.6	75.2
62	38.3	52.0	122	72.7	76.0
64	39.5	52.8	124	73.9	76.8
66	40.6	53.6	126	75.0	77.6
68	41.8	54.4	128	76.1	78.4
70	42.9	55.2	130	77.2	79.2

Chart 2

Height (cm)	Men Age in Years										
	20	25	30	35	40	45	50	55	60	65	70
150	25.6	24.2	22.8	21.4	20.0	18.6	17.2	15.8	14.4	13.0	11.6
155	26.6	25.2	23.8	22.4	21.0	19.6	18.2	16.8	15.4	14.0	12.6
160	27.7	26.3	24.9	23.5	22.1	20.7	19.3	17.9	16.5	15.1	13.7
165	28.7	27.3	25.9	24.5	23.1	21.7	20.3	18.9	17.5	16.1	14.7
170	29.8	28.4	27.0	25.6	24.2	22.8	21.4	20.0	18.6	17.2	15.8
175	30.8	29.4	28.0	26.6	25.2	23.8	22.4	21.0	19.6	18.2	16.8
180	31.9	30.4	29.1	27.6	26.2	24.8	23.4	22.0	20.6	19.2	17.8
185	32.9	31.5	30.1	28.7	27.3	25.9	24.5	23.1	21.7	20.3	18.9
190	34.0	32.5	31.2	29.7	28.3	26.9	25.5	24.1	22.7	21.3	19.9
195	35.0	33.6	32.2	30.8	29.4	28.0	26.6	25.2	23.8	22.4	21.0
200	36.1	34.6	33.2	31.8	30.4	29.0	27.6	26.2	24.8	23.4	22.0

Height (cm)	Women Age in Years										
	20	25	30	35	40	45	50	55	60	65	70
150	7.7	6.7	5.7	4.7	3.8	2.8	1.8	0.9	0.0	−1.0	−2.0
155	8.1	7.1	6.1	5.1	4.2	3.2	2.2	1.2	0.2	−0.7	−1.7
160	8.5	7.5	6.5	5.5	4.5	3.6	2.6	1.6	0.6	−0.3	−1.3
165	8.8	7.8	6.9	5.9	4.9	4.0	3.0	2.0	1.0	0.0	−0.9
170	9.2	8.2	7.3	6.3	5.3	4.3	3.4	2.4	1.4	0.5	−0.5
175	9.6	8.6	7.6	6.7	5.7	4.7	3.7	2.8	1.8	0.8	−0.2
180	10.0	9.0	8.0	7.0	6.1	5.1	4.1	3.2	2.2	1.2	0.2
185	10.4	9.4	8.4	7.5	6.5	5.5	4.5	3.5	2.6	1.6	0.6
190	10.8	9.8	8.8	7.8	6.8	5.9	4.9	3.9	3.0	2.0	1.0
195	11.2	10.2	9.2	8.2	7.2	6.2	5.3	4.3	3.3	2.4	1.4
200	11.5	10.5	9.6	8.6	7.6	6.7	5.7	4.7	3.7	2.7	1.8

oxygen intake rises extremely rapidly at first and then more slowly, before finally attaining a plateau; this remains constant over the total work time (Fig 40). After the end of work, there is almost a mirror image of events at the onset of exercise. The oxygen intake returns not linearly, but exponentially to its initial level. An analogous pattern to oxygen intake can be seen for carbon dioxide production. At the onset of work, there is an oxygen debt which remains constant over the total working period and is paid off after the end of work.

These relationships are different during heavy physical work (Fig 41). In this case, the oxygen intake continually increases. Although the steepness of the rise is flatter, no plateau is attained because the work must be interrupted due to exhaustion. The oxygen requirement is thus

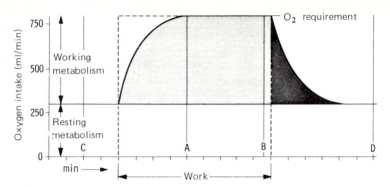

Fig 40   Contracting and repaying oxygen debt during mild work. Oxygen debt remains constant during steady-state exercise. From Lehmann G.: Praktische Arbeitsphysiologie, 2nd ed. (Stuttgart: Thieme 1962).

higher than the maximal oxygen intake. In these two examples, there are obvious differences in the pattern of oxygen intake, depending on whether the work is mild or severe. In the first case, after a certain time there is the attainment of an equilibrium (also called the steady-state) between the oxygen intake and the oxygen required by the

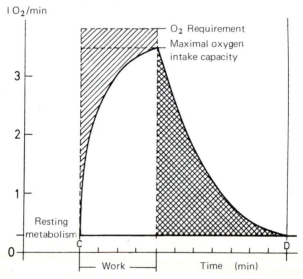

Fig 41   Oxygen intake and oxygen debt during intense exercise. Oxygen debt increases over entire work period.

musculature. In this range, the work continues aerobically, i. e., the oxygen transported to the musculature is sufficient. In the case of the higher work load, the need for oxygen cannot be covered and the musculature goes into an oxygen debt. As we already know, this oxygen deficit corresponds to both the utilization of creatine phosphate and the formation of lactic acid. The lactic acid is resynthesized to glucose or oxidized at other sites of the body; while this does occur during exercise, most is converted during the recovery period. As a result, the actual oxygen intake is not an indication of the metabolism at any moment. During mild and severe exercise, the body goes into an oxygen debt. Although this remains constant over the entire period of mild exercise, it continually increases during heavy work until exhaustion.

The maximal oxygen debt that the body can contract is 15–20 l oxygen. Of this, about 4 l can be attributed to the alactacid energy production involving the reserves of energy-rich phosphates. The remainder is the result of an accumulation of lactic acid (lactacid O_2 debt).

Taking the maximal oxygen debt into consideration, the concentration of lactic acid in the blood can be as high as 150 mg/100 ml.

The most appropriate way to determine the oxygen requirement and thus the energy metabolism for a given work load is related to the amount of work to be performed. For aerobic work, steady-state values can be used. The expired air is collected over a portion of the steady-state period, as shown in Fig 40 by lines A and B. The volume of this expired air is determined and its oxygen content is analyzed, so that the oxygen intake for this portion can be determined in the usual manner.

In the case of heavy work or a fluctuating work load, the integral method is used, i. e., all of the expired air must be collected from the onset of work (line C of Fig 41) to the time when the oxygen requirement has returned to its resting level (line D). From the total oxygen intake and the total carbon dioxide production, it is possible to compute the total metabolism during exercise. If the resting metabolism is then subtracted from the total metabolism, the result is the working metabolism for that exercise; this is the product of work load and work time.

1.9.3.2 Efficiency

Efficiency is a measure of how economically a work load is performed. During the discussion of energy-producing processes for physical work, it was determined that muscle is able to transfer the energy contained in ATP to the contractile proteins. The energy needed for ATP restitution comes from the oxidation of foodstuffs. In order to study the economy of this chain between foodstuff oxidation and

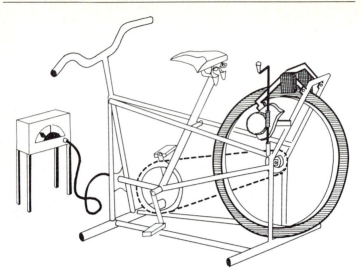

Fig 42   Bicycle ergometer with drum brakes, as designed by E. A. Müller. Rear wheel is supplied with copper disk, which intersects lines of force of slotted magnet. The deeper the magnet is lowered, the more the wheel will be braked. The work load can be adjusted by the small crank and is the result of both the depth of the magnet and the speed of wheel rotation. The speed of wheel rotation is maintained by means of a pacemaker (left side of picture).

mechanically performed work, it is necessary to determine the amount of actual work performed and the working metabolism needed.

To measure work or performance, there are a series of ergometers; the most commonly used of these is the bicycle ergometer (Fig 42). With this, the subject turns a flywheel at a known number of pedal revolutions. The flywheel is externally braked by a magnetic, electromagnetic or friction brake. The work performed by the subject is the product of the wheel velocity and the resistance applied to the wheel. The same principle of work load measurement is also applied to crank ergometers; in this case, the subject cranks against a braking force at a known velocity.

It is also possible to stress subjects in a defined manner by letting them run on a treadmill, where the speed and angle of incline are exactly adjusted (Figure 43). In principle, such a treadmill is made of two rollers and an endless rubber belt operated by a motor. The subject runs in a direction opposite to that of the belt, thus remaining in the same place. The work (W) performed by the subject is therefore:

$$W = M \ v \ \sin \alpha.$$ (23)

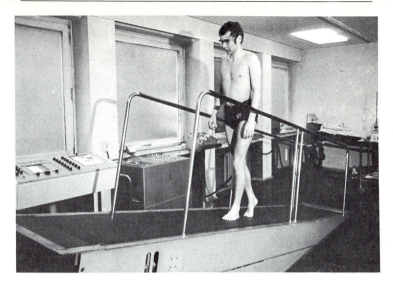

Fig 43   Treadmill ergometer. An endless band is moved in opposite direction of running subject at a defined grade, such that subject remains in same position. With this kind of ergometer, work performed (kpm/second) is: W = M v sin α, where M is body weight of the subject, v is the velocity in m/second and α is per cent grade of treadmill.

where α is the angle of incline, v is the velocity of the treadmill and M is the body weight of the subject.

Ergometers can be standardized in either kpm/second or in Watts (1 kpm/second = 9.81 W; 1 W = 0.102 kpm/second).

Metabolism and work expressed in physical terms are linked together by efficiency, which is stated as a per cent of the mechanical heat equivalent:

$$427 \text{ kpm} = 1 \text{ kcal } (100\% \text{ efficiency}) \qquad (24)$$

The rate of efficiency is thus:

$$\eta = \frac{\text{Work performed} \times 100}{\text{Work metabolism} \times 427} \% \left[\frac{\text{kpm} \times \text{minute} \times \text{kcal}}{\text{minute} \times \text{kcal} \times \text{kpm}} \right]. \qquad (25)$$

If one measured work metabolism for the same physiologic work performance on the bicycle ergometer and the treadmill, he would find that it is different. The same amount of physical work does not require the same metabolic work. Work metabolism depends much more on the type of work. If the muscle contracts isometrically, then the product of force × distance is 0, i. e., the external work is 0, even though an increased amount of energy is expended. The rate of efficiency of an isometric contraction (static holding work) is thus zero. If the muscle contracts against a force of zero, then the external work and the corresponding efficiency are also zero. Between these two extremes, efficiency increases to reach its optimal value at a certain ratio of force and distance. Efficiency determined in this way is called net efficiency. It is also possible to measure gross efficiency. This is obtained when total metabolism is inserted into formula 25 instead of work metabolism. In this case, resting metabolism must also be added to the work metabolism.

The ratio between force and distance is not the only determinant of efficiency. In addition, the time factor has a major role but is not included in the physical definition of work. Thus, efficiency is low when a high level of work is performed over a short time or, vice versa, when a low level of work is carried out over a long period of time. This is true even though the product, performance × time = work is the same in each case.

If one proceeds on the assumption that performance equals force times velocity and then alters only the force at each predetermined speed of movement, net efficiency remains constant over a wide range (Johannson rule). The rate of efficiency is altered only when accessory muscle groups must be applied to overcome the resistance.

Because the same amount of physical work can cause different metabolic effects, the amount of energy required is the decisive factor in the evaluation of a work load and never the actual work performed. Figure 44 gives an overview of efficiency for several practical activities. On the abscissa is the net performance, net efficiency is on the ordinate and the variable measured is energy metabolism. The highest theoretical rate of efficiency of the musculature is 30–35%; it thus appears that man measures up well to modern combustion engines. In general, the only advantage of machines is that they are adapted by engineers to their tasks; this is usually not the case with man and industrial work. For this reason, theoretical efficiency is never attained and it is possible to find types of industrial work with efficiencies of 5% or less.

A simpler and more easily understood method is to forego calculation of net and gross efficiencies, using instead calories expended per kpm as a measure of working economy. Naturally, it is also possible to use special indices in work physiology, e. g., energy metabolism per 1 m^3 of masonry, when one wants to determine which kind of stone stresses the bricklayer the least. In the same way, the energy turnover per athletic exercise can be determined when it is not possible to calculate the work done in physical terms. This method provides an easily visible measure and saves computation. The calculated relationship between gross efficiency and the efficiency quotient (cal/kpm) is shown in Table 4. In most forms of work where there has been sufficient practice (see p 264), work efficiency is independent of the level of training. Figure 45

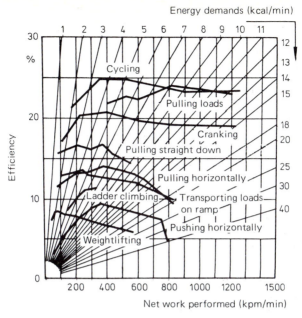

Fig 44   Efficiency of different forms of work (according to E. A. Müller).

Table 4   Conversion of Efficiency (%) into cal/kpm

Efficiency %	cal/kpm	Efficiency %	cal/kpm
1	234.19	16	14.63
2	117.09	17	13.77
3	78.06	18	13.01
4	58.54	19	12.32
5	46.83	20	11.70
6	39.03	21	11.15
7	33.45	22	10.64
8	29.27	23	10.18
9	26.02	24	9.75
10	23.41	25	9.36
11	21.29	26	9.00
12	19.51	27	8.67
13	18.01	28	8.36
14	16.72	29	8.07
15	15.61	30	7.80

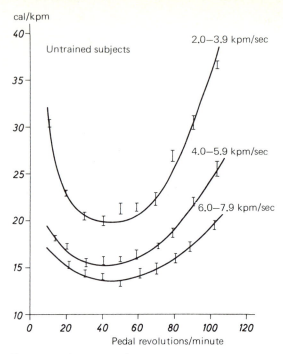

Fig 45   Energy requirement per kpm as function of pedalling frequency on bicycle ergometer at different performance ranges (untrained subjects). From Stegemann, J., Ulmer, H.-V., Heinrich K. W.: Int. Z. angew. Physiol. 25 : 224, 1968.

shows work efficiency during bicycle ergometer exercise as a function of the number of pedal revolutions; the parameter being studied is work load. It can be clearly seen that at all levels of work, optimal efficiency is found at 40–60 pedal revolutions per minute. It is markedly less with fewer or greater pedal revolutions. These results were obtained from a sample of untrained male and female subjects. Figure 46 shows the same results obtained from a highly-trained cyclist (gold medal winner in cycling at the Olympic games in Tokyo) compared to the data of untrained persons (dotted lines). As can be clearly seen, work efficiency is not affected by the level of training.

1.9.3.3 Daily Metabolism of Occupational Work

To have a gross estimation of energy requirements, the metabolism must be totaled over a certain time period. The total daily metabolism

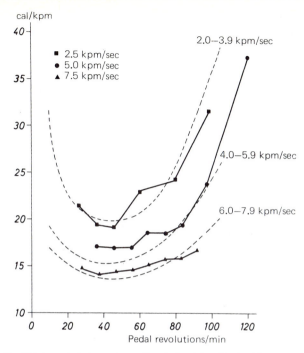

Fig 46   Efficiency is independent of level of training. Dashed lines are same as those shown in Figure 45. Heavy lines show corresponding data from an Olympic winner in cycling. From Stegemann, J., Ulmer, H.-V., Heinrich, K. W.: Int. Z. angew. Physiol. 25 : 224, 1968.

is then determined by the time of each activity multiplied by the metabolic cost of that activity. A more uniform measure of daily requirements of workers in an occupation is obtained by determining the metabolism for a whole week from its various components and multiplying this by the number of week days. Variations in energy cost found on individual days can then be used to find an average value. In this way, the total metabolism is methodically put together like a mosaic from a series of partial metabolisms, each of which must be determined according to its energy cost and duration. This is necessary if values with sufficient precision are desired.

Of course, one can also be aided by a time-motion study and a measurement of the amount of time spent in such repetitive activities as walking, sitting, standing or climbing stairs. The interindividual variation in energy costs of these activities is small and can therefore be taken from tables (e. g., those of Spitzer and Hettinger). It is assumed

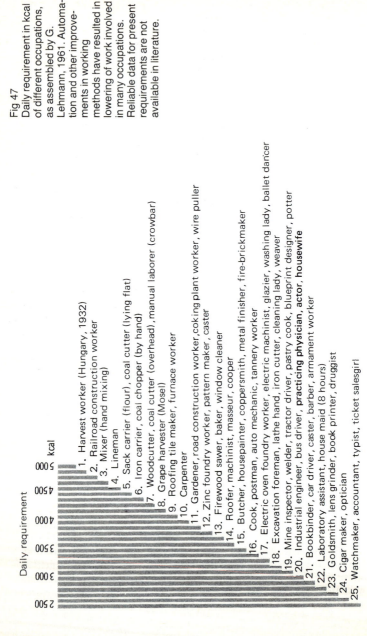

Daily requirement

kcal

5 000	
4 500	
4 000	
3 500	
3 000	
2 500	

1. Harvest worker (Hungary, 1932)
2. Railroad construction worker
3. Mixer (hand mixing)
4. Lineman
5. Sack carrier (flour), coal cutter (lying flat)
6. Iron carrier, coal chopper (by hand)
7. Woodcutter, coal cutter (overhead), manual laborer (crowbar)
8. Grape harvester (Mosel)
9. Roofing tile maker, furnace worker
10. Carpenter
11. Gardener, road construction worker, coking plant worker, wire puller
12. Zinc foundry worker, pattern maker, caster
13. Firewood sawer, baker, window cleaner
14. Roofer, machinist, masseur, cooper
15. Butcher, housepainter, coppersmith, metal finisher, fire-brickmaker
16. Cook, postman, auto mechanic, tannery worker
17. Electric oven foundry worker, electric machinist, glazier, washing lady, ballet dancer
18. Excavation foreman, lathe hand, iron cutter, cleaning lady, weaver
19. Mine inspector, welder, tractor driver, pastry cook, blueprint designer, potter
20. Industrial engineer, bus driver, **practicing physician, actor, housewife**
21. Bookbinder, car driver, caster, barber, armament worker
22. Laboratory assistant, house maid (8 hours)
23. Goldsmith, lens grinder, book printer, druggist
24. Cigar maker, optician
25. Watchmaker, accountant, typist, ticket salesgirl

Fig 47
Daily requirement in kcal of different occupations, as assembled by G. Lehmann, 1961. Automation and other improvements in working methods have resulted in lowering of work involved in many occupations. Reliable data for present requirements are not available in literature.

that an adult man requires about 2,300 kcal/24 hours for his resting metabolism and nonoccupational activity.

Measurements of energy metabolism will always be compiled for a large number of industrial activities. Of course, these energy costs are only approximate values, because working conditions can vary greatly. In this regard, one should always keep in mind that several activities which often go unnoticed (such as climbing stairs) are strenuous. For this reason, many heart patients first become aware of their sickness while doing these activities.

Figure 47 has information about the caloric costs during 8-hour occupational work in different occupational groups.

1.9.3.4 Energy Metabolism During Sport

To obtain a rough idea of the energy costs of various sports, values are given in Table 5. The greater the muscle mass used, the more asymmetric the movement and the faster the alternation between flexing and extending, then the higher the metabolic cost of each type of sport will usually be.

1.9.4 Work Loads and Their Limits

To classify energy requirements, it must be clear from the outset that a classification based on energy metabolism can only be a rough index. Other criteria will be discussed in the chapter on physical performance capacity. In this regard, it is interesting to see which energy requirements have been labeled physically mild or severe. Information on this can be found in Figure 48. On the abscissa is the range of working metabolism; total metabolism is on the ordinate. Occupational activity

Table 5   Energy Requirements of Sport Performances

Type of Sport	kcal/kg · hr
Cross-country skiing 9 km/hr	9.0
Rowing, fixed seat 6 km/hr	9.3
Running 9 km/hr	9.5
Ice skating 21 km/hr	9.9
Swimming 3 km/hr	10.7
Running 12 km/hr	10.8
Running 15 km/hr	12.1
Wrestling	12.3
Badminton	12.6
Running 17 km/hr	14.3
Cycling 43 km/hr	15.7
Cross-country skiing 15.3 km/hr	19.1
Handball (European)	19.3

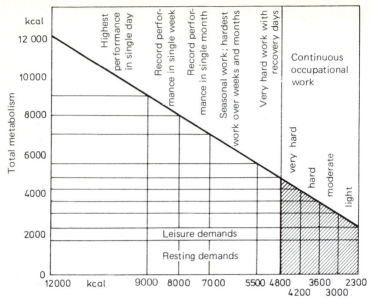

Fig 48   Daily metabolism and dietary intake. Leisure demands are about 2,300 kcal. Continuous work is possible up to about 4800 kcal/day. Highest performances in a single day can amount to 12,000 kcal. From Lehmann, G.: Energetik des arbeitenden Menschen. In: Handbuch der gesamten Arbeitsmedizin, vol. 1, Arbeitsphysiologie (Berlin: Urban and Schwarzenberg, 1961).

that can be continuously performed by man is represented by the shaded area on the right side. The limit lies somewhere near a total energy metabolism of 4,800 kcal/day; this corresponds to a work metabolism of 2,500 kcal/day. Also seen on the right side of the figure is the classification of light, moderate and heavy work in relation to metabolism. The left side shows what levels of maximal metabolism are possible during brief periods of time without causing injury, if intervals of prolonged rest are interspersed. This range is especially valid for high-level athletic performances. The limit is primarily based on the fact that with normal nutrition, no more than 4,800 kcal of nutrients per day can be supplied from foodstuffs over a prolonged period.

1.10 Nutrition and Performance

The fact that the law of conservation of energy is just as valid for the body as it is for inanimate objects means that for each unit of energy expended, a new unit of energy must be taken in if the energy content of the body is not to change. We have already seen that the unit of heat, the kilocalorie, has been generally adopted as a measure of energy loss. For this reason, the same unit is used for the computation of nutritional intake.

Nutrition is taken into the body in the form of foodstuffs; these can vary markedly depending on the customs and country. In Europe alone, it is possible to find large differences in the preferred forms of nutrition without having to travel long distances. For example, for the past 200 years the potato has been the favorite foodstuff in Germany. In Austria or Italy, people prefer noodles and other foodstuffs made with flour. Meat, milk and eggs, as well as vegetables, are appreciated all over the world as basic foodstuffs.

There are three groups of nutrients or foodstuffs. These are carbohydrate, fat and protein, each of which has its own characteristic properties.

Carbohydrates are divided into three basic groups: monosaccharides, oligosaccharides and polysaccharides. The important monosaccharide compounds are glucose and fructose (fruit sugar). Oligosaccharides are compounds formed by the combination of several units into a large molecule, e. g., two monosaccharides form a disaccharide, as is the case with lactose (milk sugar), cane sugar and beet sugar. When more than 8–10 units form a large molecule, it is labeled a polysaccharide. Examples of such polysaccharides are plant and animal starches and glycogen. The larger the number of units which are brought together in the glucosidic compound, the greater is the energy content. Details about the chemistry and biochemistry of foodstuffs can be found in textbooks of physiologic chemistry.

Fats are neutral substances. Chemically, they are the fatty acid esters of the trivalent alcohol, glycerol. Fats are predominantly found in the form of triglycerides, which usually contain 3 fatty acids. Thus, fats are complex mixtures of triglycerides. The most important, even-numbered saturated fatty acids are palmitic and stearic acid, found primarily in animal fat. The unsaturated fatty acids, oleic acid and linoleic acid, have an important role because they cannot be synthesized in the body.

Proteins are substances of high molecular weight, composed of about 20 amino acids. The individual proteins can be distinguished by the sequence of amino acids and by the number of amino acids joined together.

Several characteristics of these nutrients are important for nutrition: their energy content, their possibility of being utilized, their ease of degradation in the body and the financial costs which must be expended. In middle Europe, carbohydrates generally are about 60%, fats 20–40% and proteins 12–15% of the total amount of energy consumed. In the case of higher total energy requirements (heavy work, sport), the proportion of fats is greater, with a concomitant reduction in carbohydrates. If there are long intervals between meals, the spontaneously chosen nutrition reflects a greater proportion of fats.

1.10.1 Caloric Value of Nutrients

The physical caloric value of foodstuffs can be determined outside the body in a combustion calorimeter (Fig 49). This bomb calorimeter is a steel receptacle in which a measured amount of foodstuff is put in an atmosphere of pure O_2. The oxygen pressure is about 25 atmospheres absolute pressure. As a result, an electric spark will cause complete combustion of the substance. The heat is taken up by a water jacket insulated on all sides. The amount of heat formed is measured by the temperature increase and the specific heat of all the heated parts.

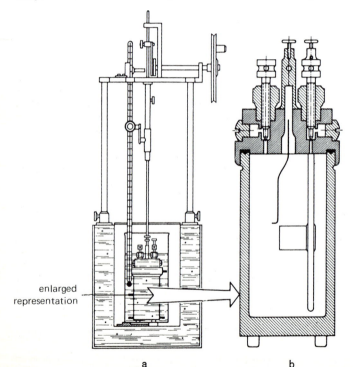

enlarged
representation

a   b

Fig 49   Bomb caloriometer, as designed by Berthelot, used for determining the caloric value of nutrients and foodstuffs. a, entire measuring device; b, combustion chamber. From Pichotka, J.: In: Kurzgefasstes Lehrbuch der Physiologie, 4th ed., ed. by Keidel, W. D. (Stuttgart: Thieme, 1975).

If a mixture of carbohydrates is combusted (this approximates its natural composition), the result is a caloric value of 4.1–4.2 kcal/gm. With a mixture of fat similar to that found in nature, values of 9.2–9.3 kcal/gm are obtained. The physical caloric value of proteins is 5.3–5.9 kcal/gm, depending on their composition. The physiologic caloric value of fats and carbohydrates in the body is the same as the physical caloric value because the same end products are produced, i. e., carbon dioxide and water. The physiologic caloric value of protein, on the other hand, shows a difference of 1.6 kcal/gm. This difference is due to the fact that the end products of the combustion process are H_2O, CO_2 and nitric oxide, whereas an end product of protein metabolism is urea, which itself contains energy. The body synthesizes this substance in order to remove the poison of the intermediate product, ammonia. With the normal composition of foodstuffs, physiologic caloric values such as those found in Table 6 are obtained.

1.10.2 Carbohydrate and Fat Requirements

The requirement for carbohydrate and fat is regulated primarily by the need for energy. Dietary protein accounts for about 300–400 kcal/day; the remainder is covered by carbohydrate and fat. In the case of nutrition which is not too unbalanced, it is of minor importance whether more or less is used of either one. Without endangering himself, man can almost completely limit his consumption of fat. However, he cannot completely replace it by carbohydrates or protein because an amount of fatty acids, the so-called essential fatty acids, cannot be manufactured from these other foodstuffs but must be ingested as fat. The essential fatty acids are necessary for the synthesis of certain biologic structures. A small consumption of fat is needed in order to receive sufficient amounts of the fat-soluble vitamins. It is also not possible to replace the total proportion of carbohydrates in the diet by fat calories. This is due to the fact that intermediate products of carbohydrate metabolism are necessary for the degradation of fats (see citric acid cycle, p 47). If they are not present, acetoacetic acid, betaoxybutyric acid and acetone bodies appear in the blood. Fat calories can account for up to 50% of the caloric requirements without the appearance of these substances in the blood. An appropriate dietary fat content is about 30% of the caloric requirements of persons who do not perform physically heavy work. For those workers who do

Table 6   Physiologic Caloric Value of Foodstuffs

Carbohydrate	4.1 kcal/gm
Fat	9.3 kcal/gm
Protein	4.1 kcal/gm

work extremely hard, 35% of the total food calories should come from fat.

It is of relatively minor importance in which form carbohydrates are consumed because starches can be relatively quickly hydrolyzed into monosaccharides in the body. Even during exercise, there is restoration of the monosaccharides which were used up. Only with prolonged work loads greater than the threshold for prolonged work (anaerobic threshold) is it recommended to give small amounts of glucose frequently, in order to avoid depleting the carbohydrate stores in the body.

With heavy physical work loads, the composition of fat is important because the velocity of absorption of fat is dependent on its melting point. For example, a high proportion of oleic acid reduces the melting point of fat to under 37C and increases its speed of absorption. Because fats are water insoluble but the lipolytic enzymes are soluble in water, the degradation of fat occurs only at the interface. Fats which are ingested in an emulsion, e. g., milk fat, can even be broken down by the relatively weak stomach lipases. The nonemulsified fats can be absorbed only after they have been in contact with the strong pancreatic lipases in the presence of bile acids.

1.10.3 Protein Requirements

Protein is the main constituent of the cell and is thus present in all the structures of the body. For this reason, the role of protein in human and animal organisms is so complex that it is impossible to give universally valid, precise details about protein requirements. A small portion of dietary protein serves as a primary source of energy when there is no surplus of the other nutrients. A further portion of the amino acids contained in protein is used in the synthesis of hormones and enzymes, to replace protein losses and for the synthesis or reconstruction of constituents in the body. Among the 20 amino acids normally available, 8 "essential" amino acids are required by man, i. e., they cannot be formed in the body but are a constituent of the body's own protein. As a result, the body needs a supply of these essential amino acids. Thus, the greater the percentage of these essential amino acids, the higher the biologic value of dietary protein. Table 7 summarizes the biologic value of several proteins. For an adult who performs only mental work, it has been estimated that the protein requirement for 24 hours is 1 gm/kg body weight. This protein is required to replenish the protein broken down during metabolism and lost through normal wear and tear. It is assumed, however, that this protein has a high biologic value, such as is found primarily in animal protein. As a rule, plant protein has only a relatively small amount of essential amino acids.

Table 7   Biologic Value of Several Proteins (according to Pichotka, full value = 100)

Protein in	Biologic value
Eggs	94
Beef	80
Milk	100
Casein	70
Oatmeal	89
Potatoes	71
Rice	68
Corn	54

This protein requirement is called the minimal balance because the ingestion and the elimination of protein are generally balanced. However, this is not the case for the growing body because new protein is constantly being formed. The minimal balance for an infant is about 2.5 gm/kg body weight per day; this decreases to 1 gm/kg body weight per day when growth has ceased. In the same sense, the needs of adults can also be increased when their muscle mass is enlarged by physical training or adaptation to work. The minimal balance not only depends on growth, but also on the active cell mass, e. g., larger amounts are found in trained heavy workers or high-performance athletes. For the same body weight, they have a 5–10% higher minimal balance.

The question of whether physical work increases the protein requirements of man has been clearly answered in recent years. Physical exercise does not increase protein metabolism when the caloric requirements during the total work time are supplied by other nutrients. The increased musculature found with training, as well as with the transition to heavy work, is dependent on the surplus of protein and the amount needed to cover the endogenous degradation of protein.

In comparison to the physically inactive worker, the heavy worker has a relatively greater active cell mass, especially in his musculature. As a result, his minimal protein balance is higher per kg body weight. These established facts suggest that those workers who do the heaviest work or high-performance athletes should have at least 1.2 gm protein per kg body weight per day. Again, this is based on the assumption that the energy requirements have been completely met by fat and carbohydrate over the entire working time. That is, the degradation of protein during work is not increased, especially when irregular meals or long intervals between them causes an increased utilization. It is important to consider these facts and to provide an even greater amount of protein when there is a deficiency due to training, after a vacation or a period of illness, and thus a greater degradation of protein. It is

Table 8 Essential Characteristics of Vitamins. From Pichotka, J., in: *Kurzgefaßtes Lehrbuch der Physiologie*, 4th ed., Ed. by Keidel, W. D.: (Stuttgart; Thieme 1975).

Name	Physiologic Importance	Experimental and Clinical Symptoms of Deficiency	Typical Human Deficiency Diseases	Presence	Recommended Daily Allowance
Vit. A = Retinol $C_{20}H_{30}O$	As vit. A-aldehyde, effective group of visual substances needed for normal epithelial proliferation	Cornification of flattened and glandular epithelium; drying hardening and peeling of the skin, cornea and mucous membrane; retarded growth	Hemeralopia; xerophthalmia; keratomalacia; hyperparakeratosis	Fish liver oil; mammalian liver; milk; as precursor (carotene) in carrots	FNB: 1.7 mg DGE: 1.7 mg
Vit. D_2 = Calciferol $C_{27}H_{44}O$ Vit. D_3 = $C_{28}H_{44}O$	Calcium resorption; calcification of bones and dentine	Inhibition of calcification; demineralization; swelling of cartilage; softening and deformation of bones	Rachitis (rickets); osteomalacia; osteoporosis	Fish liver oil; mammalian liver; animal fat; milk	L: 0.1 mg
Vit. E = Tocopherol $C_{29}H_{50}O_2$	Antioxidant	Fatty degeneration of the liver; muscle degeneration; sterility due to abortion; testicular atrophy with inhibition of spermatogenesis	None	Grain germ; peanut oil	L: 30 mg

	Function	Deficiency symptoms	Deficiency disease	Sources	Requirement
Vit. K_1 = Phylloquinone $C_{31}H_{46}O_2$ Vit. K_2 = $C_{41}H_{56}O_2$	Needed for synthesis of prothrombin	Delayed coagulation; hemorrhaging	Hemorrhagic syndrome, e.g., in the newborn	Green plants; liver	L: 1 mg
Vit. B_1 Thiamine $C_{12}H_{16}ON_4S$	As thiamine pyrophosphate, coenzyme of pyruvate oxidase, α-ketoglutarate oxidase and transketolase	Paralysis; muscle atrophy; anesthesia; myocardial insufficiency; achylia; absorption disturbances	Beriberi (polyneuritis)	Yeast; grain	L: 0.4–1.8 mg RDA: 1.5 mg EAN: 1.7 mg
Vit. B_2 Riboflavin $C_{17}H_{20}O_6N_4$	Constituent of yellow enzymes	Decreased growth rate; keratitis	None defined	Yeast; grain; egg whites; liver; milk	L: 1.6–2.6 mg RDA: 1.6 mg EAN: 1.8 mg
Vit. B_6 = Pyroxidine $C_8H_{11}O_3N$	As pyroxidal phosphate, constituent of amino acid transaminases and decarboxylases	Decreased growth rate; dermatitis; convulsions; hypochromic anemia; leukopenia	None defined	Yeast; grain; liver; meat (muscle)	FNB: 1.5–2.0 mg
Vit. B_{12} = Cyanocobalamin $C_{63}H_{90}O_{14}N_{14}PCo$	Participation in methylation, especially with nucleic acid metabolism; participation in degradation of propionic acid	Macrocytic anemia; glossitis; achylia; funicular myelosis	Pernicious anemia; Biermer's anemia	Liver; various microorganisms	FNB: 3–5 µg

Table 8 (continued)

Name	Physiologic Importance	Experimental and Clinical Symptoms of Deficiency	Typical Human Deficiency Diseases	Presence	Recommended Daily Allowance
Niacinamide $C_6H_6ON_2$	Constituent of the pyridine coenzymes (dehydrogenases)	Dermatitis of bodily parts exposed to light; stomatitis; gastroenteritis; paresthesia; dementia	Pellagra	Yeast; grain; tomatoes; liver; milk	L: 12–18 mg
Folic acid = Pteroylglutamic acid $C_{19}H_{19}O_6N_7$	Participation in the transfer of C_1 units at different oxidation steps	Megaloblastosis of bone marrow; macrocytic anemia	Pernicious anemia	Green leaves; yeast; liver; microorganisms	FNB: 0.05 mg
Biotin $C_{10}H_{16}O_3N_2S$	Constituent of carboxylation enzymes	Dermatitis; seborrhea	None	Liver; egg yolk; milk; yeast	L: 0.3 mg
Pantothenic acid	Constituent of coenzyme A	Depigmentation of hair and plumage: ataxia; paralysis: diarrhea; disturbed production of adrenocortical hormones	None	Widely available	FNE: 10 mg
Vit. C = ascorbic acid $C_6H_8O_6$	Redox system; participation in hydrolytic reactions; connective tissue metabolism	Multiple bleeding of gums, skin; susceptibility to joint infections	Scurvy	Citrus fruits; paprika; haw; parsley; black currants	L: 75 mg RDA: 75 mg EAN: 75 mg

important to remember that strength training does not produce an increase in muscle strength when there is a negative protein balance.

1.10.4 Requirements of Vitamins and Trace Elements

Diet has to supply not only caloric needs but also must contain substances which do not deliver energy themselves. If these substances are absent, pathologic deficiencies or even diseases can occur. Detailed descriptions of the structure of vitamins and the changes brought about by their absence can be found in clinical textbooks. Table 8 contains only a review of the physiologic importance, symptoms and diseases caused by a deficiency, foods in which they can be found and recommended daily allowances.

Table 9 lists the vital trace elements, with the diseases and symptoms caused by a deficiency, foods in which they can be found and daily requirements. Work and sport physiologists are especially interested in which vitamins are needed in higher doses during periods of heavy work or which vitamins can increase physical working capacity.

It has been reported that during physical stress there is an increase in the level of vitamin A in the blood. It appears questionable whether this means that there is a higher requirement for vitamin A during exercise. Doses of vitamin D do not raise physical working capacity; there is no information about an increased need during physical work.

Vitamin B_1 (thiamine) initiates the degradation of carbohydrate at the stage of pyruvic acid, since it is a constituent of cocarboxylase. The thiamine requirements in animals depend on the amount of unconverted carbohydrate. The need for vitamin B_1 is related to portions of dietary carbohydrate and protein; 1 µg thiamine is recommended for each kcal carbohydrate and protein. The improvement in endurance work capacity seen with doses of glucose does not appear if the diet contains too little thiamine. The German Nutrition Association recommends a daily dosage of 1.8 mg for physically active adults, 2.5 mg for those who do heavy work and as much as 3 mg for those who do the heaviest work or high-performance athletes.

During muscular activity, there is a marked increase in the intermediary metabolites of ascorbic acid (vitamin C); details are not known. The daily allowance of ascorbic acid should be 75 mg during severe work because a deficiency of vitamin C can diminish subjective willingness to perform. Considerable amounts of vitamins B_1 and C are excreted in sweat during work in the heat.

Among the minerals, phosphorus has a special role during active work. The excretion of phosphate by the kidneys drops to extremely low levels during work and then rises considerably afterward. There are some persons whose physical work capacity increases with doses of phosphate, while there is no effect on others. It can be assumed that

Table 9 Vital Trace Elements. From Pichotka, J. in: *Kurzgefaßtes Lehrbuch der Physiologie*, 4th ed., Ed. by Keidel, W. D.: (Stuttgart; Thieme 1975).

Element	Physiologic Importance	Experimental and Clinical Symptoms of Deficiency	Typical Human Deficiency Disease	Presence	Recommended Daily Allowance
Iron	Essential constituent of hemoglobin and different enzymes	Anemia; high-grade asthenia; alterations in skin and its appendages	Typical iron deficiency anemia; chlorosis; types of anemia (achylia, neonatorum, pregnancy)	Liver; lentils; spinach; lettuce	ca. 12 mg
Copper	Necessary for transport of iron; constituent of enzymes	Apparent iron deficiency anemia in animal experiments; disturbances of pigment formation, keratin synthesis and bone formation	Apparent iron deficiency anemia	Fish; liver; corn; beans	ca. 2 mg
Cobalt	Constituent of vit. B_{12}; activator or inhibitor of different enzymes	None (toxic in high doses: goiter formation, polycythemia vera)	None	Green vegetables; liver; kidney; meat (muscle)	–

Manganese	Activator of different enzymes; inhibitor of histidase	None with humans; depending on type of animal: anemia, sterility, hypo-thyroidism	None	Grain; root vegetables; leafy vegetables 3–9 mg
Iodine	Essential constituent of thyroid hormone	Hypothyroidism; goiter	Endemic goiter	Salt water fish; can be in drinking water; liver; eggs; spinach 100–200 µg
Zinc	Complex formation with insulin, heparin, ascorbic acid; ATP component; activator or protective substance of different enzymes	In humans, stunted growth; depending on type of animal: dermatitis, malformations, sterility	None	Liver; meat; grain; 10–15 mg fresh vegetables

there are generally sufficient levels of phosphate in the usual mixed diet to supply one's needs.

Among the trace elements cited as being necessary is iron, which is needed for the synthesis of hemoglobin, myoglobin and the cyto-chromes. In addition, iodine is mentioned because it is used to form thyroid hormone. As a rule, the intake of sodium is always assured. Under certain conditions, the ingestion of potassium is not sufficient, especially when there is inadequate intake of vegetables.

1.10.5 Diet and Performance

During and immediately after the war years, the terrible nutritional conditions gave researchers enough possibilities to study the influence of poor nutrition on physical performance. The understandable fact that physical work which exceeds the basal metabolism can be per-formed only as long as there are sufficient calories is clearly shown in Figure 50. Because workers had to unload debris from trucks, work performance was easily measured. After a certain delay, the amount of work performed proceeded in agreement with the number of calories above the basal metabolic level that the men received. At this time, the body weight rose somewhat continuously. Before the research began, the workers were underweight due to the prior conditions of undernu-trition; the average weight was 56 kg. An attempt to increase job performance through a cigarette bonus for each ton of work performed (instead of through calories) actually raised the level of work due to an increased incentive to work. This improved performance, however, was made at the expense of a weight loss.

The nutritional situation in Europe today is such that too many calories are being consumed rather than too few. In many parts of the world, hunger and its destructive effects on man's productivity is still a major factor. Hunger is not only a social problem, it also reduces man's productivity considerably.

1.10.6 Nutrient Content and Utilization of Foodstuffs

Nutrients are contained in the food we eat; they are seldom found in a pure state. Most foods are a mixture of fat, carbohydrate, protein and, most of all, water. To calculate the caloric content of foods, the considerable amount of water must be subtracted. Secondly, it is necessary to know the composition of the food. Summaries of compo-sition and energy content of foods can be found where they are sold. Use only the newest information because there are continual changes due to alterations in methods of cultivation. In addition, an individual should consider how much of the food is actually utilized in relation to his nutrient requirements. The degradation of foodstuffs is the result of the well-known digestive process. Only those substances for which

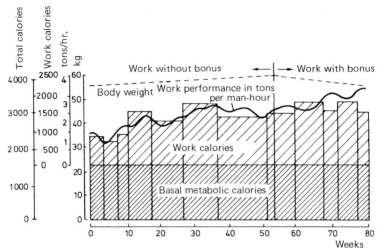

Fig 50   Relationship between work performance (continuous line) and number of calories consumed during hunger period after World War II. Work performance was increased when workers were given bonus of cigarettes, even though body weight decreased. From Kraut, H.: Ärztl. Wschr. 3 : 499, 1948.

there is a corresponding enzyme in the digestive tract can be broken down. For example, man has no enzyme system for the polysaccharide cellulose to permit the degradation of this substance. In comparison, the rat is able to almost completely survive on a diet of cellulose. Herbivorous animals are also able to digest cellulose, without having an enzyme system, through the effect of their intestinal bacteria. Not all proteins can be directly digested, particularly when they are enclosed in an indigestible plant membrane.

The ease with which food is utilized can be increased by such forms of preparation as cooking, cutting it into small bits, etc. In many cases, food from animals is more valuable in terms of its essential amino acids. The more advanced the agriculture of a country, the more preferred is an animal diet. Carbohydrates, which are produced by plants from carbon dioxide and energy from the sun, are refined in the animal's digestive system and made usable. Vegetarian diets contain many important vitamins and minerals, especially when they are eaten raw. It is always surprising when for ideological reasons, the enjoyment of meat or milk is taboo in many countries, even though the nutritional situation of the people living there frequently leads to suffering from symptoms of protein deficiency. This is one of the future tasks of aid to developing countries, i. e., breaking down unscientific dietary taboos.

2 Circulatory System and Work

We have discussed the fundamental processes of energetics and their application to physical work and sport, but have so far ignored the problem of transport. The transport of nutrients and of oxygen to the exchange site, the carrying away of end products and heat produced there and the transmission of information via the hormonal control mechanism occur through the blood. The fact that working musculature can be situated at different areas of the body places huge demands on the regulation of blood distribution. Because of numerous characteristics, blood is well-qualified to be this means of transport.

2.1 Oxygen Transport

In the section on "Energy-supplying processes for muscle activity," we established that all forms of energy supply require oxygen, when considered over a prolonged period. The oxidative processes occur in the mitochondria of the cells (p 10).

A single-celled organism living in water has no problem of supply because the oxygen pressure in the surrounding water is higher than the oxygen pressure within the cell. Oxygen always diffuses from the region of higher to the region of lower pressure. The driving force for oxygen transport is the pressure gradient. However, the cell wall of the unicellular organism represents an obstacle for oxygen and the resistance to diffusion is formed there. As an anlogy to Ohm's Law, the following diffusion equation is given:

$$\dot{V}_{O_2} = \frac{\Delta P_{O_2}}{R}, \tag{26}$$

where ΔP represents the pressure difference, R the diffusion resistance and $\dot{V}_{O_2}$ the volume of oxygen diffused per unit of time. This relationship is called the Fick Diffusion Law. The resistance to diffusion depends on the thickness of the membrane layer, the diffusion area, the solubility of oxygen and the material constants.

If there are a number of cells in combination, as is the case with man, then this simple system is no longer adequate because the thickness of the membrane layer from the skin to the body's cells is too great. It is for this reason that during the course of evolution the surrounding "ocean" was transformed into the form of blood in the body to supply the cells with the necessary oxygen.

However, if our cells were bathed only in a saline solution, as represented by the ocean, then the velocity of the circulation would have to be extremely high. This is because the solubility of oxygen in water is low, i. e., the volume of O_2 per 100 ml fluid (vol%) would be small and there would be a low transport capacity. The hemoglobin (Hb) of blood allows a small volume of blood to transport a large amount of oxygen.

By external respiration (p 142), blood is placed in equilibrium (equilibrated) with a gas mixture containing oxygen. It is replenished and transported to the tissue by the circulatory system, where oxygen is once more transferred.

Hemoglobin represents a complex protein molecule composed of globin and the actual carrier of oxygen, heme, which itself further contains bivalent iron. As is the case with all protein molecules, hemoglobin also has ampholytic traits, i. e., it can combine with both acidic and basic salts. The isoelectric point of deoxygenated hemoglobin is near the neutral point and that of oxygenated hemoglobin is in the acidic range. Thus, oxyhemoglobin (with alkaline ions) can form less salt than hemoglobin.

The transport of oxygen occurs in such a way that molecular oxygen can be stored in an easily reversible form in the hemoglobin molecule, thus forming HbO_2. When oxygen is transferred to tissue, HbO_2 is converted into reduced hemoglobin. The form that hemoglobin assumes is thus dependent on the surrounding oxygen pressure.

Hemoglobin is a constituent of the erythrocyte. Under normal conditions (sea level), the number of erythrocytes in men is about 5,000,000/mm^3 and in women about 4,500,000/mm^3. Erythrocytes swim suspended in blood plasma. The approximate levels of the constituent hemoglobin in gm/100 ml blood (gm %) in men and women are 16 gm/100 ml and 14.5 gm/100 ml, respectively. The normal range of hemoglobin attached to a single erythrocyte is between 30 and 34×10^{-12} gm. The maximal amount of oxygen that 1 gm hemoglobin can bind is 1.38 ml (as a reduced gas mixture).

2.1.1 Oxygen Dissociation Curve

To comprehend the quantitative relationship between environmental oxygen pressure and the oxygen content of blood, imagine placing a certain volume of blood in equilibrium with a certain oxygen pressure. The desired oxygen pressure is maintained when a certain mixture of oxygen and nitrogen is produced. Ideal gases follow the law that a gas exerts a fractional pressure (partial pressure), which corresponds to its fractional percentage of the volume. If the mixture is 10% O_2 and 90% N_2 at an atmospheric pressure of 760 mm Hg, the oxygen and nitrogen

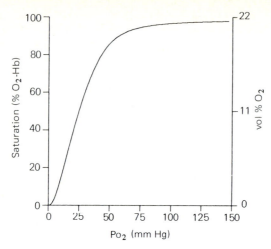

Fig 51   Standard O_2 dissociation curve calculated for pH 7.4; temperature 37 °C; PCO_2 40 mmHg; and Hb 16 gm/100 ml, as well as 2,3-DPG content in the erythrocytes of 5 mmol/L. Left ordinate shows saturation in per cent and right ordinate is the corresponding vol % O_2.

exert partial pressures of 76 mm Hg and 684 mm Hg, respectively (see also the chapter on "Respiration").

When this method is used to place the oxygen in blood in equilibrium, the following process now occurs, characterized symbolically as:

Phase 1	Phase 2	
physical solution	chemical binding	(27)
$P_{O_2} \rightleftarrows [O_2]$	$[O_2] + [Hb] \rightleftarrows [HbO_2.]$	

The fraction of HbO_2 depends on the partial pressure of oxygen (P_{O_2}). After an equilibrium has been produced, if the oxygen content of blood is determined, there is a characteristic relationship which is labeled the oxygen dissociation curve (Fig 51). It reflects the relation between the externally acting oxygen pressure (P_{O_2}) and the oxygen content of blood (vol %). In the upper range, the curve runs practically parallel to the abscissa; this means that despite the increasing O_2 pressure, no more oxygen can be taken up by the blood. All of the hemoglobin is now present in the form of oxyhemoglobin (HbO_2), i. e., the hemoglobin is saturated. To make this independent of a given level of hemoglobin, which can vary greatly from one individual to another, the point at which all hemoglobin is converted into HbO_2 is called 100% saturation. At the other end of the curve ($P_{O_2} = 0$), no O_2 is available in the blood and all the hemoglobin is present as reduced hemoglobin (0% saturation).

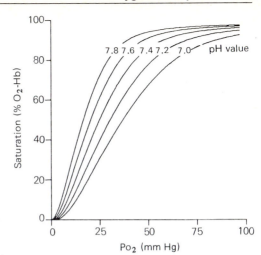

Fig 52
Relationship between
pattern of the O₂ dis-
sociation curve and
pH. Calculated for
PCO₂ of 40 mmHg,
temperature, 37 °C;
2,3 DPG, 5 mmol/L
erythrocytes.

The ordinate on the right (vol %) in Figure 51 is dependent on the
level of hemoglobin; the left ordinate is not. Since 1 gm hemoglobin
binds a reduced volume of 1.38 ml, under normal conditions and at
100% saturation this corresponds to a value of $16 \times 1.38 = 22.08$ vol %
in men and $14.5 \times 1.38 = 20.01$ vol % in women.

2.1.2 Factors Influencing the O₂ Dissociation Curve

The Danish physiologist Bohr was the first to show that the O₂
dissociation curve changed its pattern when acid was added to or
withdrawn from blood (Bohr Effect). Under physiologic conditions,
two acids produced by metabolism have an important role: volatile
H_2CO_3 (carbonic acid), which is the end product of aerobic metabolism, and lactic acid produced by anaerobic energy production. Figure
52 illustrates the influence of different levels of acidity on the O₂
dissociation curve. At rest, blood has a pH value of about 7.4 (for an
explanation of pH values, see p 333). If the blood becomes more
alkaline, the curve is clearly shifted to the left; with the addition of
more acid, it shifts to the right.
Similar shifts are initiated by varying the CO₂ pressure. Proceeding
from a normal P_{CO_2} of about 40 mm Hg, such as prevails at rest, the
curve shifts to the left with a lower P_{CO_2} and to the right with higher
levels of P_{CO_2} (Fig 53).
The third important influence on the O₂ dissociation curve is temperature. The higher the blood temperature, the further to the right the
dissociation curve is shifted. Acids, as well as temperature, displace

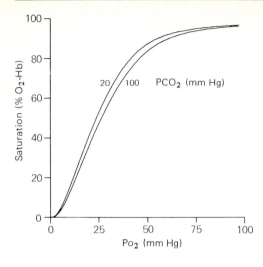

Fig. 53
Relationship between
O_2 dissociation curve
and PCO_2. The pH is
7.4; temperature,
37 °C; 2,3 DPG,
5 mmol/L erythro-
cytes. Curve is
affected only slightly
by PCO_2 when pH is
held constant.

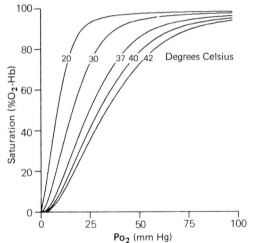

Fig 54
Relationship between
O_2 dissociation curve
and blood tempera-
ture; pH, 7.4; PCO_2,
40 mmHg; 2.3-DPG,
5 mmol/L erythro-
cytes.

the balance between hemoglobin and HbO_2 in a characteristic direction (Fig 54).
Another more recently discovered influence on the O_2 dissociation
curve is exerted by a substance which is produced indirectly by the
glycolytic metabolism of erythrocytes, 2,3-diphosphoglycerate (2,3-
DPG). The greater the concentration of 2,3-DPG in erythrocytes, the
further to the right the dissociation curve is shifted (Fig 55).

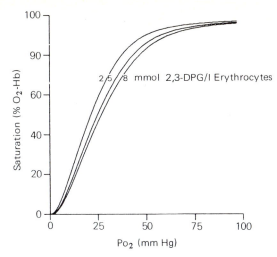

Fig 55 Relationship between O_2 dissociation curve and erythrocyte content of 2,3-DPG. Calculated for PCO_2, 40 mmHg; pH of 7.4 and temperature of 37 °C.

Displacements of the O_2 dissociation curve to the right and left have a practical significance for the performance of physical work. An athlete who is "warmed-up," for example, has more favorable P_{O_2} conditions in the periphery.

To understand this process, we must remember that the volume of oxygen that diffuses into the muscle cell per unit of time is regulated by the O_2 pressure gradient between the extracellular space and the site of utilization, the mitochondria, where the oxygen pressure is extremely low. The greater the work load, the greater must be the volume of oxygen per unit of time that reaches the cell because the metabolism increases with increasing work loads. If the supply of oxygen is insufficient, then anaerobic energy production begins; this results in fatigue. As an example, imagine that there is an oxygen saturation of 60% and a temperature of 30 C at the level of the muscle capillaries, such as is normally present at rest (p 188). Using Figure 54, if we pass a perpendicular line from the point of intersection between the line for 60% saturation and the curve for 30 C to the abscissa, we find that there is a P_{O_2} of 25 mm Hg. If under the same conditions, we warm up the muscle to 43 C, as is often the case during high work loads, then the P_{O_2} is almost 50 mm Hg. While maintaining the same saturation, the effect has been to double the pressure gradient.

Similar considerations are also valid for P_{CO_2} and pH. At high work loads, more CO_2 and more hydrogen ions appear in capillary blood, thereby raising the pressure and the supply of O_2 to the cell.

The uptake of O_2 at the lungs is facilitated by the opposite effect, i. e., colder air improves saturation at the same pressure. This is especially seen with work loads at altitude, at which time the oxygen pressure in the inspired air is extremely low.

2.2 Transport of Carbon Dioxide

Carbon dioxide is the end product of aerobic metabolism. Analogous to oxygen transport, a simple physical solution would not be adequate for the transport of carbon dioxide with the circulating blood volume. This is true even though CO_2 dissolves about 25 times better than oxygen in water.

When CO_2 leaves the cell, it dissolves immediately into the tissue fluid. A part of the CO_2 becomes hydrated, i. e., with water it forms carbonic acid,

$$CO_2 + H_2O \rightleftarrows H^+ + HCO_3^-, \tag{28}$$

which subsequently is partially dissociated into H^+ and HCO_3^- ions, according to the laws of dissociation. The ratio of dissociated to undissociated carbonic acid is described by the first apparent dissociation constant K':

$$K' = \frac{[H^+]\ [HCO_3^-]}{[H_2CO_3]}. \tag{29}$$

If the equation is solved for H^+, the Henderson-Hasselbach equation is obtained:

$$[H^+] = K' \cdot \frac{[H_2CO_3]}{[HCO_3^-]}. \tag{30}$$

Because it is not possible to decide exactly how large the ratio of dissolved to hydrated CO_2 is, it is assumed that all the CO_2 is hydrated and thus K' is designated the first apparent dissociation constant.

Carbon dioxide also becomes diffused in relation to the pressure gradient, thus dissolving in blood plasma, as well as in erythrocytes. There is an enzyme in erythrocytes called carboanhydrase, which accelerates this hydration. For this reason, H_2CO_3 appears there with a much greater velocity than in the plasma.

In regard to its ionic composition, the erythrocyte is constructed in a manner similar to that of tissue cells (p 10). In erythrocytes, K^+ ions and large negative protein anions predominate; in this case, protein is primarily represented by hemoglobin. Because protein ions are ampholytic in nature, the bond between protein ions and K^+ ions can be considered as being the salt of a weak acid. Salts of weak acids always react with a strong acid in such a way that the salt of the stronger acid and the free weaker acid result; the balance is regulated by the concentration.

When H_2CO_3 is thus formed in erythrocytes, it reacts with the protein in the following way:

$$K^+ + Prot^- + H_2CO_3 \rightleftarrows H\,Prot + K^+ + HCO_3^-. \qquad (31)$$

As a result, the salt $KHCO_3$ appears; it is called either potassium bicarbonate or potassium hydrogen carbonate. With the formation of this salt, osmotic pressure in the erythrocytes rises, causing an inflow of water.

The result of this is that Cl^- ions now enter into the erythrocytes from plasma. There is thus an anion exchange between Cl^- and HCO_3^-, so that there will be a balanced CO_2-dependent concentration of $NaHCO_3$ in plasma.

The most important transport substance for CO_2 is sodium bicarbonate, allowing erythrocytes to have a transport function. Because the direction of the reaction described here is dependent on CO_2 pressure, there will be a certain equilibrium regulated for each CO_2 pressure.

2.2.1 CO₂ Dissociation Curve

The quantitative relationship between CO_2 content and CO_2 pressure is reflected in the CO_2 dissociation curve (Fig 56), which is obtained in the same way as the O_2 dissociation curve. The upper curve depicts the total content of CO_2 in vol% as a function of P_{CO_2}. The lower linear function represents the fraction of "dissolved" H_2CO_3. The amount of ideal gases in solution is proportional to the pressure; as such, the angle of rise at a given temperature is specific for each gas. This angle is called the Bunsen absorption coefficient (Table 15). It can be clearly seen that the upper right portion of the curve is almost asymptotic to a parallel of the physically dissolved H_2CO_3. This means that apparently no new sodium bicarbonate can be formed at this level and that the total content can increase further only in solution. The reason for this is that all the available alkali liberated by protein binding has been converted into sodium bicarbonate.

2.2.2 Factors Influencing the CO₂ Dissociation Curve

In connection with equation 31, the equilibrium between protein acids on the one side and the carbonic acid and bicarbonate on the other side depends on the acid strength, under otherwise equal conditions. However, the chemical measure for this strength is its dissociation constant. At the same CO_2 pressure, the concentration of bicarbonate present in erythrocytes (and thus also the plasma concentration) depends on the ratio of the dissociation constants of the two participating acids. Because the dissociation constant of carbonic acid is constant, that of protein (hemoglobin) depends on the oxygen being

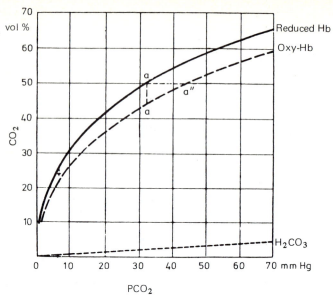

Fig 56 CO_2 binding curve of blood. From Schneider, M.: Einführung in die Physiologie des Menschen (Berlin: Springer, 1966).

carried. Oxyhemoglobin is a stronger acid than reduced hemoglobin. As a result, more bicarbonate can be formed with reduced hemoglobin, because the equilibrium of equation 31 is shifted to the right. This fact is seen in Figure 56, in which the CO_2 dissociation curves of reduced and oxidized blood are depicted. The effect of O_2 saturation on the CO_2 dissociation curve is called the Haldane Effect. Its importance for CO_2 binding at the tissue, especially during work, and CO_2 release at the lungs is as follows. Without any real change in CO_2 pressure, more or less volume of CO_2 can be transported when the O_2 saturation of the blood varies at the same time. The influence of non volatile (fixed) acids (e. g., lactic acid) on the CO_2 dissociation curve will be discussed in the section on "Respiration."

2.2.3 Review of CO_2 Transport

Figure 57 summarizes total CO_2 transport under resting conditions. In addition to the mechanism of binding in the form of bicarbonate discussed earlier (at a P_{CO_2} of 40 mm Hg, this amounts to about 44 vol %), approximately 3 vol % is physically dissolved and another 3 vol % is present as carbaminohemoglobin. Carbaminohemoglobin is the direct combination of H_2CO_3 with hemoglobin, even though the

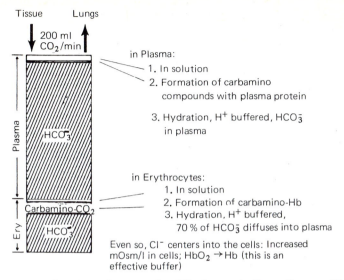

Fig 57 Overview of means for CO_2 transport. From Ganong, W. F.: Medizinische Physiologie (Berlin: Springer, 1971).

binding does not occur on the heme portion, as is the case with oxygen, but on the globin component.

2.3 Transport of Heat, Nutrients and Fixed Metabolic End Products

Blood has approximately the same heat capacity as water. This is in the physiologic range 1 kcal/per degree centigrade; a metabolic rate of 1 kcal/minute with a blood flow of 1 L/minute produces an additional arteriovenous temperature difference of 1 C.

The major combustible of muscle work, glucose, is present in arterial blood at a concentration of about 100 mg/100 ml; it is regulated hormonally and is independent of dietary intake or utilization. The concentration of fats and fatty acids is strongly influenced by dietary intake.

Lactic acid is one of the more important fixed metabolic end products. Its concentration in venous blood at rest is about 1 mmol/l; this can rise to a maximal level of 2.5 mmol/l with exhausting exercise. The concentration in arterial blood is usually lower because the heart and resting muscles can use lactic acid as a fuel.

2.4 Blood Flow and Muscle Supply

Before undertaking a closer analysis of blood flow regulation, it should be clear how the level of metabolism in a muscle group affects the arteriovenous difference (AVD) of various measurable variables when we conceptually vary blood flow (Fig 58). First, we assume that the total metabolism of a working muscle group is 1 kcal/minute and that this level of metabolism remains constant over the total work period. Assume further that arterial blood has a normal saturation, hemoglobin content and other parameters and that the subject is at sea level. If blood flow to the muscle group is 1 l/minute, then we can determine the individual variables. According to the data on page 61 and at a metabolism of 1 kcal/minute, the muscle group uses about 200 ml O_2 per minute and produces 170 ml CO_2 per minute at an R of 0.85. The maximal amount of oxygen that one liter of blood can transport is 200 ml, so that almost all the oxygen in the blood will be withdrawn. In this case, the AVD for oxygen is 20 vol % or 100% saturation and the AVD for oxygen pressure is about 100 mm Hg, i. e., the oxygen pressure in the vein leaving the muscle is almost 0. At the same time, CO_2 content increases from 48 vol % in arterial blood to 65 vol %. Considering the fact that the blood is completely reduced at the same time, venous P_{CO_2} rises to about 70 mm Hg, resulting in an AVD for P_{CO_2} of 30 mm Hg. The AVD for temperature in this case would be

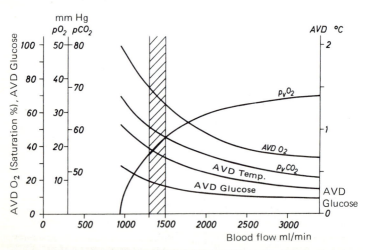

Fig 58   Relationship between arterial venous difference (AVD) of various factors and blood flow of muscle group having a metabolism of 1 kcal/minute. For AVD curve for glucose, it is assumed that muscle does not utilize muscle glycogen but takes its energy only from glucose delivered via blood.

exactly 1 C. If it is additionally assumed that the muscle group takes its energy from blood glucose and not from muscle glycogen, then the AVD for glucose is about 25 mg/100 ml. Figure 58 illustrates AVDs and venous pressures, calculated in the same way, for a series of blood flow levels. The figure makes it clear how closely blood flow must follow metabolism in order to maintain the muscle's ability to work. The mechanisms involved in the optimal adjustments of local and systemic circulation to metabolism will be discussed in the next section.

2.4.1 Adjustment of Local Muscle Blood Flow (Control System)

Mean blood pressure is altered by about 20–50% during exercise. It is largely kept constant by a control system (p 111). In comparison, blood flow (i. e., the volume flowing through musculature per unit of time) can be modified by about 1,200%. Muscle has at its disposal a control system which is capable of measuring out blood flow, such that it is as little as possible but as much as needed to fulfill metabolic needs.

Unfortunately, all authors do not use the same nomenclature for the terminal vessels. It is known that arteries branch out into the "smallest arteries," which then pass into the "arterioles." The smallest arteries still have a complete smooth layer of muscle, whereas the arterioles represent the transition to capillaries and no longer have a complete muscle layer. The capillaries are only simple endothelial tubes. Earlier, the smallest arteries were also called arterioles, often causing confusion in the literature.

The corrective devices (Suppl, p 302) are the smallest arteries, whose smooth musculature is predominantly innervated by sympathetic or vasomotor fibers. With stimulation of the vasomotor fibers, smooth muscle fibers contract and thereby diminish the radius of the blood vessel. Relaxation of vascular muscle primarily occurs passively due to pressure.

The simplest way of demonstrating the relationship between flow rate $(dV/dt = \dot{V})$ and the morphological parameters is by using the Hagen-Poiseuille Law:

$$\dot{V} = \frac{\pi \, r^4 \, \Delta P}{k \, \eta \, l}. \tag{32}$$

This states that in a rigid-walled tube with a radius r and a length l, flow rate $(\dot{V})$ is proportional to the 4th power of the radius and to the pressure difference (ΔP) at the beginning and end of the tube and inversely proportional to the length (l) and viscosity (η). In this case, k is a dimension constant. With this law, the effect of blood vessel width on blood flow can be approximated. A dilatation of the smallest artery by a factor of 2 (a doubling of the radius) produces an increase in blood flow by the factor of $2^4 = 16$. Furthermore, because pressure gradient has only a linear effect, it participates little in the blood flow change discussed here.

Arteries are not rigid tubes, however. They are elastic blood vessels which become dilated with increasing inner pressure, as is the case with an elastic rubber hose. Thus, blood flow also depends on the magnitude of the absolute blood pressure. The "autoregulation" of the vascular muscles has a counterbalancing effect on this. There are distinct differences in this musculature in the various sections of the circulatory system. The term autoregulation means that smooth vascular musculature contracts when it is stretched and without the involvement of the sympathetic nervous system. As a result, perfusion is automatically regulated, as is shown in Figure 59. If the perfusion pressure increases, the result is an initial disproportionate rise in flow due to the elastic stretching of the vessel wall. At a certain pressure, autoregulation counteracts this and perfusion remains constant. Autoregulation of muscle is most important under resting conditions. It also has extensive control over blood flow to the kidneys and brain.

There is also the problem of viscosity (flow resistance) because the Hagen-Poiseuille Law is valid only for the flow of homogeneous fluids (also called "Newton flow"). In the case of blood, there is "Maxwell flow" because the corpuscular elements (blood cells) move in the fluid. In addition, erythrocytes have a tendency to agglutinate. As a result, shearing forces are active in the terminal vessels to separate them from each other. The viscosity measured in a viscosimeter would thereby be apparently greater. There is thus an "apparent" viscosity, which varies greatly in the different sections of the circulatory system.

2.4.2 Problem of Adequate Stimulation for Blood Flow Control

Figure 60 is a schematic diagram of a control loop. First of all, let us look at the familiar controlled process (C), i. e., the area affected by the regulator (R). Each modification of regulator output causes a

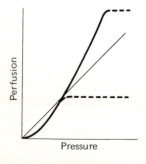

Fig 59 Schematic representation of pressure-perfusion relationship of blood vessel. Linear function reflects dependency of perfusion according to Hagen-Poisseuille Law. Parabolic function shows that relationship is nonlinear in elastic vessels because there are changes in radius with increasing internal pressure. Dotted portion reflects effect of autoregulation on vessel areas with differing sensitivities. From Schneider, M.: Einführung in die Physiologie des Menschen (Berlin: Springer, 1971).

change in the effector, i. e., the width of the smallest artery. How is the regulator controlled? Although the mechanisms have been intensively investigated for the past 100 years, they are still partially disputed.

It is certain that blood flow is held in check by the influence of the sympathetic vasomotor system. It can be seen in Figure 60 that a positive influence (increased tonus) on the usual negative polarity produces a decrease (–) in blood flow. Because under normal conditions the sympathetic system is always tonically active, blood flow at rest is always low. With the addition of exercise, a second influence becomes active and counteracts the sympathetic system. Phenomenologically, it appears that the local influence of the sympathetic system is inhibited by factors associated with exercise. In the control diagram, this is symbolically represented by the additive effect of the sympathetic system (+) and exercise (–).

There is still disagreement about which substances cause these exercise effects. We will not go into the various contradicting theories but limit ourselves to those hypotheses which seem most probable at the present time.

Because these influences are coupled with the production of energy, they must appear in the extracellular space and then neutralize the effect of the sympathetic system. One of these influences is probably K^+ ions, which originate from the cell. Vasodilatation is produced by perfusing potassium compounds into a muscle group that has been isolated from its circulatory supply.

Higher potassium concentrations have been found in venous blood leaving an exercising muscle than at rest. It is probable that potassium is not completely pumped back after it is released during the depolarization process. An increase in hydrogen ion concentration in the extracellular fluid effectively produces an exit of K^+ ions from and an

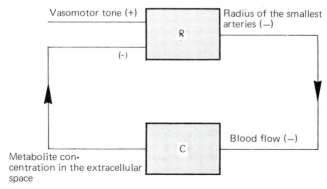

Fig 60   Cybernetic diagram of blood flow regulation.

entrance of Na^+ ions into the cell. Evidently, an elevated H^+ concentration inhibits the functioning of the sodium-potassium pump. In addition to potassium concentration in the extracellular space, osmotic pressure, hydrogen ion concentration in extracellular fluid and a reduction in oxygen pressure have additional roles in the control of vessel width. Thus, we are probably talking about a multifactorial control system. The effect of the appearance of increased levels of inorganic phosphate on blood flow regulation is in doubt.

2.5 Cardiovascular System

From an exercise physiology standpoint, we have put muscle and its blood flow at the beginning. Only when there is sufficient blood in the arterial system under adequate pressure, however, can the control system depicted in the last section become effective. Because the cardiovascular system has such an important role in exercise physiology, we must spend more time and effort studying it.

The human heart can be thought of as a double pump, with a "left heart," which supplies the systemic circulation, and a "right heart," which pumps blood to the pulmonary circulation.

Each of these halves of the heart has an antechamber (atrium) and a chamber (ventricle). The atria receive blood from the veins. Between the atria and the ventricle are the atrioventricular valves. The left ventricle is connected to the aorta and the right ventricle to the pulmonary artery. The aortic valves are situated between the ventricle and the arterial system. The contraction of the heart, a hollow muscle, plus the function of its valves, produces the desired blood flow. A schematic view of the circulation can be seen in Figure 61. The general circulation is composed of a number of parallel, interconnected circulatory sections which supply the individual organs. The fractional resting blood flow of the individual sections is given as a per cent of cardiac output. The arterial system has relatively strong and rigid walls, which produce a high level of resistance, and goes from the left side of the heart to the capillaries in individual organs. The remainder of the system, with its relatively large volume and elastic walls, is called the low-pressure or capacitance system.

2.5.1 Basic Characteristics of the Myocardium

The mechanisms for excitation and contraction of the heart are similar to those of skeletal muscle (p 14). The resting potential of the myocardium is about -80 mV. Individual fibers are separated by cell membranes but react as one large cell. Action potentials exhibit a rapid rise but a slow repolarization phase (Fig 5, f, p 13). During electromechanical coupling, the calcium ions enter the cell directly

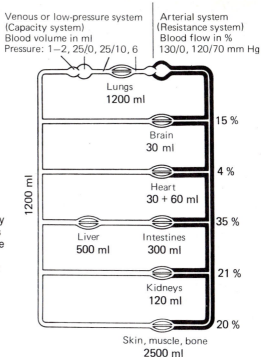

Fig 61
Diagram of circulatory system. Percentages on right side of picture show that portion of cardiac output at rest which flows to different organs. From Schneider, M.: Einführung in die Physiologie des Menschen (Berlin: Springer, 1971).

Venous or low-pressure system (Capacity system)
Blood volume in ml
Pressure: 1—2, 25/0, 25/10, 6

Arterial system (Resistance system)
Blood flow in %
130/0, 120/70 mm Hg

Lungs
1200 ml

Brain
30 ml — 15 %

Heart
30 + 60 ml — 4 %

Liver   Intestines
500 ml   300 ml — 35 %

Kidneys
120 ml — 21 %

Skin, muscle, bone
2500 ml — 20 %

1200 ml

from the interstitium, i. e., they are not contained in special vesicles, as in skeletal muscle. As a consequence of this slow repolarization phase, the myocardium cannot be tetanized.

Several myocardium cells have a particularly unstable membrane potential. These "pacemaker" cells depolarize rhythmically and have a faster repolarization phase than the other myocardial cells. The stimuli which are propagated from these cells determine the frequency of myocardial contractions. The most important pacemaker is the sinus node, which exhibits the highest spontaneous frequency. From there, the excitation spreads over the atrial musculature to the atrioventricular node. Pacemaker cells are also found here but they have a slower spontaneous frequency; this node becomes active only if the sinus node fails. From the atrioventricular node, the excitation is conducted over His's bundle and the rapidly-conducting Purkinje fibers to all parts of the ventricle.

Under special conditions, almost all parts of the myocardium are capable of spontaneous depolarization. The spontaneous frequency decreases from the base to the apex of the ventricle, however.

2.5.2 Myocardial Supply

The myocardium is supplied by the coronary arteries. Regulation of blood flow is similar to that of working skeletal muscle. The myocardium uses almost all the oxygen in the blood, even under resting conditions. In the case of large cardiac outputs, the increased need must be predominantly covered by an increase in blood flow to the coronary arteries. A reduction in O_2 pressure in the blood is a particularly strong stimulus for increased blood flow.

2.5.3 Action of Myocardial Nerves

Heart rate, as well as the force of contraction of the myocardium, is influenced by the myocardial nerves. The heart is supplied by the sympathetic nervous system (cardiac nerves), which has the transmitter substance norepinephrine. Norepinephrine increases the frequency of spontaneous discharge, producing a higher pacemaker frequency (positive chronotropy). The action potential is also prolonged, allowing an increased influx of calcium and thereby a higher contractile force (positive inotropy). In addition, norepinephrine causes a shortening of the excitation transmission (positive dromotropy) and a reduction of the stimulus threshold (positive bathmotropy). These effects usually increase myocardial performance.

The parasympathetic nervous system (vagus nerve) primarily affects the atrium, acting here as an antagonist to the sympathetic system. Heart rate is lowered essentially via the effect of the transmitter substance, acetylcholine.

2.5.4 Electrocardiogram

The electrocardiogram (ECG) represents the algebraic sum of myocardial action potentials. Because of the good electric conductivity of tissue, the ECG can be obtained from different parts of the body surface and recorded as the potential difference. Usually, it is electronically amplified and then recorded. Electrocardiography has been developed as a diagnostic method and today is a special science. We will discuss only some of the basic facts needed in exercise physiology.

With the standard limb lead I (electrodes on the right and left arms), the electrodes are connected to the recording apparatus so that the pen will deflect upward when the left arm is positive. In lead II (electrodes on the right arm and left leg), there will be an upward deflection when the leg is positive; the same is true for lead III (electrodes on the left arm and left leg).

A typical ECG waveform is reproduced in Figure 62. Because at the present time it is possible to relate the basic electrophysiologic events

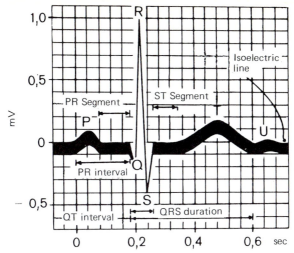

Fig 62 Typical ECG pattern. From Ganong, W. F.: Medizinische Physiologie (Berlin: Springer, 1971).

to the form of the curve to only a limited extent, the different waves are given alphabetic names. The P wave is associated with the stimulation of the atrium; the Q, R, and S waves are called the ventricular complex.

Because it is not necessary for the exercise physiologist to make clinical diagnoses, we will not be concerned with the pathologic waveforms of the ECG. Within the scope of exercise physiology, the ECG permits recording or counting the heart rate using the appropriate apparatus. It makes it possible to determine the regularity of the heart beat; this is of importance in the evaluation of the capacity to work and to exercise. There is a general rule that irregular intervals between the individual R waves at rest, in otherwise clinically healthy persons, suggest a good level of training. The greater the performance requirement, the more regular the intervals become. Irregular wave intervals (arrhythmias) during exercise suggest a pathologic condition. The exercise test should be interrupted and the individual should then see a physician.

When using the same lead, the relative height of the T wave, in comparison to the other waves, is generally considered a good indicator of the level of training of the heart. In the case of a well-trained athlete, if the T wave becomes smaller during the course of training, this suggests the presence of overtraining. The T wave becomes smaller when there is a relative oxygen deficiency. Because all systems (muscle fiber thickness, mitochondria, capillaries) do not adapt at the same rate during cardiovascular training, the heart can be temporarily in a relative oxygen deficiency during exercise. When this happens, it is recommended that training be interrupted for several days until the T waves have again reached their initial size.

In the case of highly-trained athletes, dissociation of ventricular and atrial rhythm (heart block) occasionally occurs at rest. The influence of the vagus is so great in these trained people that the pacemaker frequency of the sinus node becomes extremely low and the substitute pacemaker of the atrioventricular node spontaneously takes over. During exercise, myocardial function is immediately normalized. For this reason, these functional blocks are not dangerous.

2.5.5 The Heart as a Pump

Because of the special importance of the heart for physical activity, the classic pressure-volume diagram obtained in animal research will be briefly discussed to assist the understanding of the effects of exercise and training. First of all, regard Figure 63, A. When a pharmacologically paralyzed heart, with its nervous system still intact, is passively filled with blood, the greater the filling of the heart, the more pressure must be applied. The relationship between volume and pressure is called the passive pressure volume characteristic (compliance of the resting heart). If the left ventricle is stimulated to contract against completely closed valves, a characteristic peak pressure will be developed for a given level of filling (*perpendicular arrows*); this pressure will differ for each initial level of filling. If a line is drawn connecting all these end points, then the curve of the "maximal isovolumic pressures" (MIP) or isovolumic maxima will be obtained. When the use of appropriate equipment makes it possible for the heart to ejects its volume without an increase in pressure (isobaric), then the corresponding maximal ejection volumes for different initial levels of filling (*horizontal arrows*) are obtained; these "minimal residual volumes" remain in the left ventricle. The connected line of the end points is the curve of "minimal isobaric volumes" (MIV). Both curves, the isovolumic maxima and the isobaric minima, are curves of the heart which are not found under physiologic conditions.

The results depicted in Figure 63, B resemble the physiologic pressure pattern and permit the construction of the afterloaded isobaric contraction curve. Beginning with a given initial level of filling, after a stimulation the pressure will be isometrically increased while the volume remains the same, i. e., it is isovolumic. When tension reaches the desired level, the contraction is isobarically completed. From a practical point of view, these conditions can be obtained using a safety valve with an adjustable pressure resistance that has been placed behind the aortic valve.

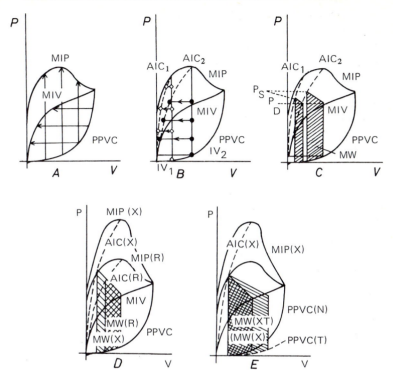

Fig 63 Heart pressure-volume curves. A, pattern of minimal isobaric volumes (MIV) and of maximal isovolumic pressures (MIP) of isolated mammalian hearts. B, construction of afterloaded isobaric contraction curves (AIC) of isolated mammalian hearts at varying initial volumes (IV). C, diagram of myocardial work (MW) of mammalian heart at varying levels of filling. D, effect of exercise on myocardial work (MW). E, effect of training on myocardial work. P, pressure; V, volume; P_S, systolic pressure; P_D, diastolic pressure; R, at rest; X, effect of exercise; N, nontrained; T, effect of training; PPVC, passive pressure volume characteristics.

In the two extreme cases (where the pressure resistance of the safety valve = 0 and = ∞), a point on the isovolumic and isobaric maxima curves is found. A different residual volume is found in between these two extremes, depending on the pressure at which the isometric contraction is finished and the isobaric contraction occurs. When all the end points are interconnected, the afterloaded isobaric contraction curve (AIC_1) is obtained which corresponds to the initial volume (IV_1). If another initial volume (IV_2) is chosen, another afterloaded isobaric contraction curve (AIC_2) will be obtained.

Figure 63, C characterizes the physiologic pressure pattern. Starting with a given initial level of filling, the ventricle first contracts isometrically. As a result, the pressure is increased until it is equal to the diastolic pressure (P_D) in the aorta. As soon as the ventricular pressure surpasses P_D, the aortic valves open. In this way, a portion of the blood in the ventricle is ejected following a further rise in pressure, until the afterloaded isobaric contraction curve for the given initial level of ventricular filling is reached. At rest, the rule is that the ventricle is filled to such an extent that about one half of the intraventricular volume is ejected (stroke volume to residual volume = 1 : 1).

Under the influence of the sympathetic system during exercise, the afterloaded isobaric contraction curve is steeper (Fig 63 D). This means that at the same initial level of filling, the heart ejects a greater volume. Stroke volume is increased at the expense of residual volume. As a result, the heart becomes smaller with each contraction; this finding has also been confirmed using x-rays.

The effect of training on the myocardium can be seen in Figure 63, E. In the trained heart (athletic heart), the resting tension curve is flatter and residual volume is greater. During exercise, this larger residual volume can be mobilized.

To completely explain the increased cardiac output during exercise, consideration of the equilibrium curves alone is not sufficient. An elevation of sympathetic tone in the heart would remain essentially ineffective if there were not a simultaneous reduction in peripheral resistance. There is a limited blood depot in the lungs which would be rapidly depleted if more blood were not brought back from the periphery to the lungs via the right side of the heart. Under physiologic conditions, the exact amount of increased blood needed by the heart is returned to it during mild, steady-state exercise. As a result, a sufficient level of filling is guaranteed. During the ejection period of each systole, there is a strong suction effect on the venous system, such that the level of the closed valve moves toward the functionally fixed apex of the heart ("valve level mechanism"). The effect of this is like an injection pump, i. e., during the relaxation period, the valve opens and spills the accumulated blood volume into the ventricle. In this way, the heart is rapidly refilled. Except at high heart rates, during which the duration of the diastole is particularly shortened, the heart always maintains a sufficient filling volume.

During strenuous physical work with small muscle groups or during static holding work, when the blood requirement and blood flow are not matched (p 254), the heart rate is too high for the required work load. At this work level, too little blood returns to the heart, resulting in an insufficient filling during diastole and thus a smaller stroke volume.

Therefore, under physiologic conditions, filling volume and not filling time is the parameter which primarily limits cardiac output.

2.6 Blood Pressure Control in the Arterial System

The control loop for blood pressure regulation has four sensors, two of which are symmetrically placed at the bifurcations of the common carotid arteries into the internal and external carotid arteries (carotid sinus bodies). Two other sensors lie in the wall of the aortic arch (aortic bodies). The carotid sinus bodies are connected to the circulatory center of the medulla oblongata via carotid sinus nerves; the aortic bodies are connected to it via the vagus nerves and several connections. The circulatory center can be functionally divided into a heart center and a vasomotor center. From here, the correcting devices of the control loop (effectors) will be regulated. When cardiac output is varied, it can alter inflow to the arterial system. The resistance to flow is regulated centrally by the vasomotor center and locally by products of metabolism. From the standpoint of theoretical control, the pressor receptor reflex represents the regulator (R) and the vascular system, which the regulator affects, represents the controlled process (C).

2.6.1 Characteristics of the Controlled Process

The controlled process is characterized by an elastic circulatory system which splits into ever smaller vascular branches. The central part of the arterial system consists predominantly of elastic material, but also shows a small plastic component. The peripheral portion of the system also contains vascular musculature which is capable of actively altering the width of the vessels. The peripheral drainage of the arterial system is constituted by the so-called "smallest arteries" because it is their vascular musculature which regulates local peripheral resistance. These vascular muscles are supplied by the sympathetic nervous system. Excitation of the sympathetic system causes a contraction of the muscles of the smallest arteries and thus an elevation in flow resistance. The reciprocal interactions between the central influence of the sympathetic system and local factors have already been discussed on page 103. Total peripheral resistance is formed by the individual resistances according to the equation:

$$\frac{1}{\text{Total R}} = \frac{1}{\text{R 1}} + \frac{1}{\text{R 2}} + \ldots \frac{1}{\text{R n}}. \tag{33}$$

Individual resistance of a smallest artery follows the Hagen-Poiseuille Law:

$$R = \frac{k \cdot \eta \cdot l}{\pi \, r^4}. \tag{34}$$

In physics, pressure is defined as the force per unit of surface. It is usual in medicine to state pressure in millimeters of mercury, whereby 760 mm Hg = 1 kp/cm^2 = 1 atmosphere = 10 m water column. The comparative pressure for blood pressure is always environmental atmospheric pressure.

Arterial pressure (P) is the result of two factors: momentary blood flow intensity at the initial portion of the aorta ($\dot{V}$) and total peripheral resistance (R). According to Ohm's Law, both variables are linked together and multiplied. Thus, for blood pressure:

$$P = \dot{V} \cdot R. \tag{35}$$

What makes it difficult to comprehend the pressure patterns in the arterial portion of the circulation is the fact that cardiac output is ejected into the arterial system in pulsating waves. During each ejection period, the arterial wall stretches and transforms a part of the kinetic energy into potential energy due to its elastic characteristics. During diastole, the potential energy is once again transformed into kinetic energy. If the arterial system did not possess this "Windkessel" (air chamber) function, the momentary flow volume $\dot{V}$ would drop to 0 during diastole. According to formula 35, blood pressure P would also drop to 0. During systole, there would be extremely high values, depending on the maximal value of $\dot{V}$. Blood pressure and blood flow are stabilized via the "Windkessel" function of the large vessels. At rest, blood pressure varies with each heart beat between a minimal value (diastolic pressure) and a maximal value (systolic pressure). The difference between the two is called blood pressure amplitude or pulse pressure.

It is possible to draw a series of conclusions from the values of systolic, diastolic and pulse pressure. For this purpose, imagine a simplified aorta as an elastic tube with a narrowing at the end which represents peripheral resistance.

1. In this model, if stroke volume, heart rate and peripheral resistance are held constant and only the distensibility of the aorta is altered, then the greater the distensibility, the smaller the pulse pressure. At the same time, mean pressure remains constant.

2. If at the same distensibility and a constant heart rate and stroke volume, only peripheral resistance R is increased, then the mean pressure in the equation will increase. In this case, diastolic pressure rises more than systolic pressure. More blood will now be stored in the "Windkessel" during systole because less can flow out during the ejection period due to the elevated resistance. During diastole, the more highly filled "Windkessel" empties itself more slowly because of

the high resistance, so that diastolic pressure at the onset of the next systole is even higher than normal. The increased resistance and larger volume in the "Windkessel" thus potentiate their effect on diastolic pressure.

3. Blood pressure also increases when only stroke volume is higher. Because the additional fraction of stroke volume now primarily fills the "Windkessel" and its outflow is not hindered during diastole, this rise in pressure is noticed especially in the systolic pressure.

4. When the other variables are held constant, an increased heart rate causes a symmetric rise in both systolic and diastolic pressure.

In summary, mean pressure can always be determined by the product of cardiac output and peripheral resistance. The related systolic and diastolic pressures, as well as pulse pressure, depend a great deal on the per cent distribution of stroke volume between that portion which flows out during the ejection period in excess of the peripheral resistance and that portion which is stored in the "Windkessel" during this time.

2.6.2 Characteristics of the Sensor

The receptor fields are imbedded in the arterial wall. As an approximation, the La Place Law is valid here:

$$\sigma = \frac{r \cdot p \cdot k}{d}, \tag{36}$$

whereby σ is wall tension, p is inner pressure of the vessel, d is vessel wall thickness and k is a proportionality constant. Deviations arise primarily because wall thickness in the area of the receptors is not constant. For the total range of information, there is a characteristic S-shaped wave in a statically measured pressure range of about 70–200 mm Hg.

The receptors show a proportional-differential relationship, i. e., they react to a stepwise change in pressure with a response function of the second order (Suppl, p 304). In the physiologic pressure range of 80–150 mm Hg, the response function is assymmetric ("unidirectional rate sensitivity"). This means that pulsating pressures in the carotid sinus will always provoke a stronger response than the equivalent static pressures.

2.6.3 Behavior of Open and Closed Control Loops

In research on dogs, static and dynamic behavior of the open control loop can be determined with sufficient precision. The artificial perfusion of the circulation with and without control gave the results

Table 10   Loop Gain V_O and Control Factor R_B for Different Combinations of Receptor Areas

Receptor Area*	V_O	R_B
R	0.4	0.715
RL	0.62	0.617
RLAA	0.8	0.556

* R, only right carotid sinus intact; RL, right and left carotid sinus intact; RLAA, all receptor areas intact.

displayed in Table 10. The control factor is obtained from the relationship:

$$R_B = \frac{\text{Steady-state with control}}{\text{Steady-state without control}}. \qquad (37)$$

The amplification factor is:

$$V_O = \frac{1}{1 + R_B}. \qquad (38)$$

The influence of the four receptor fields is not additive but there is a quasimaximal value formation, i. e., the sensor which has the highest pressure takes over about 80% of the control. The influence of the receptor fields is ideally represented in Figure 64.

The dynamics of the system can be seen in the polar plot of the frequency response (Fig 65). With its help, it is possible to state whether the control loop being studied may become unstable and, if

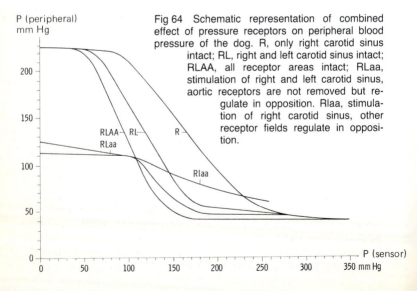

Fig 64 Schematic representation of combined effect of pressure receptors on peripheral blood pressure of the dog. R, only right carotid sinus intact; RL, right and left carotid sinus intact; RLAA, all receptor areas intact; RLaa, stimulation of right and left carotid sinus, aortic receptors are not removed but regulate in opposition. Rlaa, stimulation of right carotid sinus, other receptor fields regulate in opposition.

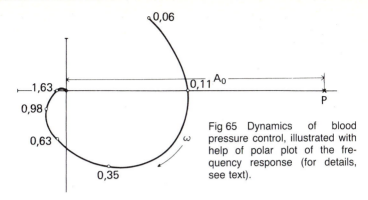

Fig 65 Dynamics of blood pressure control, illustrated with help of polar plot of the frequency response (for details, see text).

so, at what frequency. The stability criterion of Nyquist is used (p 307), with which critical point P_c is sought by using the diagram of the polar plot. This is found when the initial amplitude A_0 (in the present example, 26.5 mm Hg) is incorporated on the negative real axis. The critical point P_c is not shown in Figure 65. It lies on the same ordinate as point P. Control loops are also stable under closing conditions if the critical point P_c is to the left of the polar frequency response locus to which an observer drifts after increasing the frequency. The further the point P_c is from this locus, the more stable is the control process.

If the present control loop is evidently stable, then under physiologic conditions, there are four pressor receptor areas which are active and not a single carotid sinus, as shown here. Because the "loop gain" factor V_0 is doubled, it is to be expected that the response amplitude will also be exactly doubled. Thus, there is no instability. The natural frequency of the control loop is 140 mHz; this corresponds to a periodic duration of about 7 seconds. This frequency is in the range of what has been called THM waves (p 118). The spontaneously appearing 7-second rhythm in blood pressure is probably an expression of the control deviations.

2.6.4 Effectiveness of Control

The more a control loop can compensate for its disturbance input (load), the more effectively it works. The effectiveness of control under static conditions is relatively low.

The dynamic control factor can be determined as a function of the frequency of a sinusoidal disturbance input from the polar fequency response locus of the interrupted control loop, in which the distance from a definite frequency point on the locus to point P is measured and its value divided by the value of the initial amplitude A_0. If this measurement is done with several frequency points and the calculated

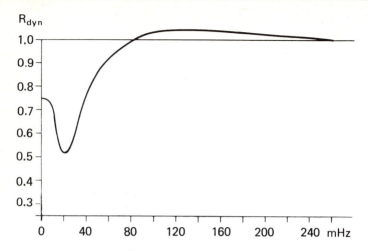

Fig 66   Dynamic control factor as function of sinusoidal stimulus frequency of the carotid sinus (in 10^{-3} Hz). The smaller the control factor, the better is the adjustment of the disturbance. $R_{dyn} = 1$ signifies an uncontrolled condition; $R_{dyn} > 1$ means a deterioration of control compared to unadjusted condition (experiment on a dog).

values connected with one another, then the dynamic control factor R_{dyn} is obtained as a function of the interfering frequency (Fig 66). A control factor of 0 means that the total interference is completely regulated. A control factor of 1 means that there is an uncontrolled condition and a control factor greater than 1 means that the interference is amplified by the effect of the regulator. According to Figure 66, the control effect is reversed under the influence of frequencies lasting between 13 and 4 seconds, i. e., it is slightly worse than the uncontrolled condition. In man, only characteristics of physiologically pulsating pressures have been obtained currently. In this case, the amplification factor is approximately the same as that found in the dog, but it can be influenced by physical training. A comparison of the mean characteristics of trained and untrained persons can be seen in Figure 67. As a basic rule, the better the level of training, the worse the quality of control (p 117). In man, the characteristic is basically nonlinear and much more effective in compensating for a drop in blood pressure than for a rise in pressure. The differences between animals and man apparently can be related to the adaptation to the upright position.

The influx into the arterial system and the level of peripheral resistance are variables which are adjusted by the pressure in the arterial system,

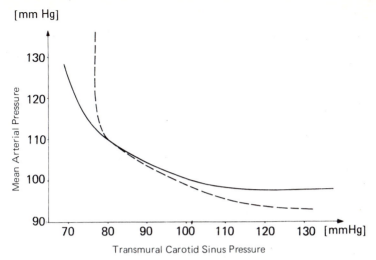

Fig 67   Mean arterial blood pressure as function of transmural pressure of carotid sinus (solid line, endurance-trained athlete; dotted line, untrained person). In man, control against drop in pressure is more effective than against rise in pressure. In addition, sensitivity of system is reduced in trained persons.

in the same way as a regulatory process. The system is especially effective in regulating an acute drop in blood pressure. Mean blood pressure is a constant manipulated variable which needs to be maintained; according to the equation, it is the product of cardiac output and total peripheral resistance. If there is a reduction in blood pressure at the level of the baroreceptors in the carotid sinus or at the aortic receptors (e. g., due to a change in body position or to a loss of blood), then there will always be a reduction in the inhibition of the sympathetic tone, which is normally under the inhibitory control of the blood pressure regulators. There are baroreceptors in the carotid sinus and at the aortic arch whose adequate stimulus is represented by vasodilatation. The greater the vasodilatation, the stronger will be the stimulation of the associated nerve. In the circulatory center, the information is then negatively polarized, so that the afferent stimulation produces an efferent inhibition of sympathetic tone and vice versa. Therefore, with a drop in blood pressure, the sympathetic system will influence the heart (increased frequency and contractile force of the myocardium) and the smallest peripheral arteries, whose wall musculature contracts due to the secretion of norepinephrine at the sympathetic nerve endings. Within the range of control, sympathetic tone is increased until the moment when the product of cardiac output and peripheral resistance reaches the new set point.

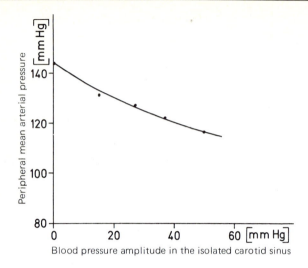

Fig 68   Peripheral arterial mean pressure as function of blood pressure amplitude at constant mean pressure of 105 mmHg in isolated dog carotid sinus. From Stegemann, J.: Med. Klinik 64: 1375, 1969.

In comparison to technical control systems, the function of blood pressure regulation is sluggish and imprecise. For this reason, blood pressure at rest is continually fluctuating around its set point. In addition to the oscillations which take place in synchronization with the heart beat, there are a number of slow blood pressure waves. Respiratory oscillations in blood pressure are primarily explained today by the fact that the pressor receptor reflexes are inhibited during inspiration and facilitated during expiration. There are still slower Traube-Hering-Meyer (THM) waves with a periodic duration of about 7 seconds or longer. Under normal circumstances, these waves can modify the systolic, diastolic and mean blood pressures by as much as ±10 mm Hg.

Blood pressure amplitude has an important effect on the adjustment of mean arterial pressure. If a carotid sinus of a dog is isolated by ligating all of the arteries leaving this area (thus producing a dead end), it can be clearly seen how different pressure amplitude stimuli in this dead end can affect mean blood pressure (Fig 68). In this case, an alteration in pressure was produced in the dead end such that its mean pressure was 105 mm Hg and its blood pressure amplitudes corresponded to changes of 20, 40 and 60 mm Hg. Peripheral mean pressure dropped with increasing blood pressure amplitude.

2.6.5 Effect of Blood Pressure on Blood Vessels

With the help of the La Place Law (p 113), it is possible to correctly comprehend the detrimental effects of prolonged blood pressure eleva- tion on blood vessels. The vessel wall is more markedly stretched due to the higher pressure and because of an elastic wall structure, the radius will also be larger. As a result, vessel tension also increases. Each pressure elevation in the vessel produces a potentiated rise in vessel wall tension. As in all tissue which is chronically overstrained, the vessel wall becomes hypertrophied. The greater its transverse wall tension, the more it becomes hypertrophied. As a result of the wall thickness enlargement associated with wall hypertrophy, wall tension is primarily compensated for by an elevated internal pressure; this is gained at the expense of a narrowing of the internal diameter and distensibility. The reduction in internal diameter leads to a rise in peripheral resistance of the smallest arteries, because their normal wall thickness is small compared to their diameter (Table 11). The rise in resistance produces an elevated diastolic pressure and thus a higher mean pressure. The effect of this rise in blood pressure then becomes intensified via the same mechanism, causing further narrowing of the peripheral circulatory system. This vicious circle is supported by the reduction in distensibility of the large vessels. In this way, a constantly elevated blood pressure (hypertension) is developed. As long as the vessel wall hypertrophy has not surpassed a certain value, it is rever- sible.

2.6.6 Resting Blood Pressure as a Function of Age

Resting blood pressure is that pressure obtained at rest without emotion or fear on the part of the subject. In a person with a labile circulatory system, blood pressure can rise due to the stimulation associated with the measurement. In addition, the effect of exercise on heart rate, especially in untrained persons, can continue for up to 2 hours. Resting blood pressure is also related to age; the normal physiologic variation is shown in Figure 69. The figure shows clearly

Table 11   Diameter and Wall Thickness at Different Regions of Arterial System (from Burton, A. C.: Physiol. Rev. 34: 619, 1954).

	Diameter (mm)	Wall Thickness (mm)
Aorta	25	2
Average arteries	4	1
Smallest arteries	$30 \cdot 10^{-3}$	$20 \cdot 10^{-3}$
Precapillary sphincter	$35 \cdot 10^{-3}$	$30 \cdot 10^{-3}$

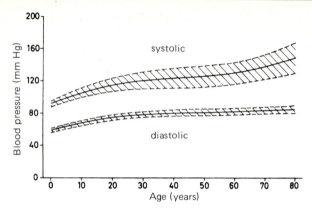

Fig 69   Normal resting blood pressure and its physiologic range as a function of age. From Stein, E. in: Lehrbuch der Inneren Medizin, ed. by Gross-Jahn (Stuttgart: Schattauer, 1966).

that the old rule of thumb that systolic pressure should be 100 plus age is not valid for all ages. A diastolic pressure measured at actual rest which is over 95 mm Hg is generally pathologic. However, a systolic pressure of 150 mm Hg in older persons can be completely normal.

2.7 Regulation of Blood Volume

The blood volume of an adult, untrained person is around 5 l. In endurance-trained athletes, it can be up to 15% higher. Normally, about 85% of the blood is located in the venous capacitance system, with the rest found in the arterial system. Blood is in water balance with the extravascular space, especially at the capillaries, because the amount of water in blood is adjusted according to osmotic and hemodynamic pressure gradients. The capillary wall is impermeable to protein but permeable to water and salts. It reacts as a semipermeable membrane, with the result that the osmotic pressure of blood is about 30 mm Hg higher than that of extracellular fluid. Because this pressure difference is related to the difference in protein concentrations, it is called the colloid osmotic or oncotic pressure. The cell membrane is largely impermeable to salt, so that the distribution of fluid between the extracellular and intracellular space is regulated by the osmotic pressure gradient between cellular fluid and its surrounding fluid. Blood volume cannot be considered in isolation. Each disturbance of osmotic balance, whether it be due to changes in the amount of water, protein or mineral, produces an alteration in blood volume, normally held constant by a control system. Disturbance inputs for this system

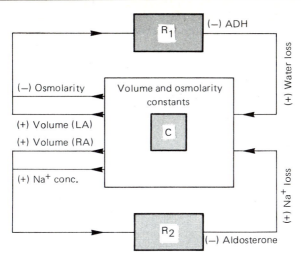

Fig 70   Diagram of regulation of blood volume and of osmotic pressure (for details, see text). LA, left atrium; RA right atrium.

include intermittent fluid intake, water resulting from the metabolism and sweat, which is an important variable influencing the level of both water and minerals. People who work in the heat (see heat regulation) can lose up to 6 l of fluid per shift. The correction device of the control loop is the kidney, which can alter its excretion of water and salt.

Blood volume is predominantly adjusted by a control system (Fig 70) whose sensor is chiefly located in the left atrium and reacts to distension. The greater the atrial filling, the more activated it becomes.

The information is transferred via the nervous system to the hypophysis, where it controls the release of antidiuretic hormone (ADH). This reflex is called the Gauer-Henry reflex after its discoverers. When the volume in the intrathoracic space increases, atrial receptor activity becomes greater and the secretion of ADH is diminished.

When ADH secretion is reduced, water excretion by the kidney increases. There is then an increased urine elimination, until such time as the fluid volume is once more normalized. The sensors in this control loop are the volume receptors, its correction device is the water elimination by the kidney and the controlled process is circulatory filling. The control loop also has a second sensor which reacts to changes in osmotic pressure. If osmotic pressure drops, the correction device (kidney) eliminates more water via the same ADH mechanism. A second independent control loop which affects the same control system regulates the secretion of a second hormone, aldosterone, produced in the adrenal cortex. The volume receptors of this control loop are primarily in the right atrium. When filling increases, aldosterone activity drops and the elimination of sodium increases. In the same manner, an increase in plasma sodium concentration is regulated via another sensor. A rise in blood osmotic pressure causes an additional feeling of thirst and thus a craving for water to drink.

2.8 Adjustment of the Circulatory System to Work

The adjustment of the circulatory system to work is characterized by a quantified elevation in sympathetic tone; it is quantified because the magnitude of sympathetic tone within a given area of control is exactly adapted to the level of metabolism in the musculature and thus to its needs. Within the area of control, the principle is as follows. Sympathetic tone within the whole body is elevated. Consequently, peripheral resistance increases in all those areas where it is not hindered from doing so by exercise reactions (p 103). It is known that the threshold for the sympathetic system is higher in working muscles. Thus, the blood becomes redistributed, i. e., blood flow to resting tissue is limited as much as possible, while it is elevated in working muscle (collateral vasoconstriction). If the muscle performs rhythmic (dynamic) work, blood flow is even assisted by the muscle pump. At the same time, the sympathetic tone of the heart is raised. This causes an increased cardiac output because more blood is returned to the heart.

The question is raised as to how the magnitude of sympathetic tone can be determined with sufficient precision to be able to draw conclusions about the area of control, which apparently is also the area related to physical working capacity. Because man cannot continuously measure epinephrine and norepinephrine and then only at great expense, he must be satisfied with a relative measure. An example of such a measure is the continuous determination of heart rate (also called pulse rate), first because it is simple to measure, but even more so because it causes few repercussions. Few repercussions in this case means that an isolated variation in heart rate has essentially no effect on cardiac output during steady-state.

This fact may be difficult for mechanically minded persons to understand because they compare the heart to a pump with a constant stroke volume which has been conveyed from a large reservoir. In reality, the heart can convey only the venous return. If heart rate is higher but there is a constant peripheral resistance and a constant cardiac output, then stroke volume would be reduced. There are investigations on men having to live with artificial pacemakers, the rate of which can be externally controlled. During constant exercise, cardiac output remains constant in these people, even when heart rate is changed.

When the teleological reason for the adjustment of heart rate is considered, the conclusion is that it is less responsible for gradations in cardiac output than for the economic promotion of cardiac output. As in skeletal muscle, efficiency of the myocardium depends on an optimal force-velocity relationship. It appears that heart rate helps keep myocardial efficiency at its high level so that the cardiac output which is available from the peripheral resistance can be pumped with the lowest possible oxygen requirement.

Therefore, few repercussions also means that an isolated change in heart rate hardly affects cardiac output or its related variable, blood pressure, which via its pressor receptors would again counterregulate sympathetic tone. For this reason, heart rate during exercise is an optimal measure of the sympathetic activity initiated by muscle receptors.

2.8.1 Behavior of Heart Rate during Work

As we have already seen in the discussion of metabolism, work is performed either aerobically or anaerobically (p 49). "Unlimited" amounts of aerobic work can be done (the word unlimited naturally refers only to muscular factors). To penetrate more deeply into the mechanism of heart rate regulation during exercise, the behavior of heart rate during aerobic and partially anaerobic work above the threshold for prolonged work should be differentiated. As a practical measure, the threshold for prolonged work is defined as that amount of work which can be performed for up to 8 hours (corresponding to the usual duration of a shift) without symptoms of muscle fatigue. During the transition to light physical work under the threshold for prolonged work, heart rate increases and reaches a constant value after 2–3 minutes; this heart rate can usually be maintained for the duration of exercise. The final value obtained in this way is proportional to the oxygen intake. If the work load is increased in a stepwise manner, then a definite work load is reached, at which heart rate leaves the linear steady-state relationship to the oxygen consumption. Although the oxygen debt entered into at the beginning of work (p 65) remains constant over the total work time in this range, the time of performance is limited. After a certain time, it must be interrupted due to fatigue.

This level is called the apparent steady state. During this apparent steady state, heart rate can be maintained at its constant value for at least 3–4 hours. Only after this time is there a sudden rise, which signals the arrival of impending exhaustion.

If there is a further progressive increase in work load, the heart rate no longer reaches a steady state (even at the onset of exercise) but continually increases until the work must be interrupted due to exhaustion. The greater the difference between the actual amount of work being done and the threshold for prolonged work, the steeper will be the angle of rise in heart rate.

The energy metabolism needed for the work load, and not the work load itself, is the deciding factor for heart rate adjustment, as well as for the threshold for prolonged work. It follows that during negative work (braking work), even though the physical amount of work performed is the same, the metabolism is lower; both the threshold for prolonged work and heart rate adjustment have correspondingly lower values. If the same amount of physical work is performed on the bicycle ergometer at different pedalling rates, there are differences in efficiency. In this case also, heart rate follows the metabolic level. When a given amount of work is carried out by a larger number of muscle groups, requiring a given metabolic level which is under the threshold for prolonged work, then as expected, heart rate will drop to its proper steady-state value. If the muscle mass involved in the work is reduced, heart rate then rises continually. The smaller the muscle mass used to accomplish a given work load, the steeper the rise in heart rate and the sooner the work must be interrupted because of exhaustion.

If an untrained person performs the same submaximal work load either on a bicycle ergometer with the leg muscles or on a crank ergometer with the arms, then there will be about the same rise in heart rate for the same oxygen requirement. If the subject now trains daily at high work loads on the bicycle ergometer, the heart rate after training will be higher at the same work load and same oxygen intake when the work is performed with the untrained arm musculature than with the trained leg muscles.

A special kind of work is the so-called static holding work (isometric exercise). It is characterized by the fact that muscle blood flow can increase only a small amount due to compression of the vessels. As a result, there is an imbalance between oxygen supply and oxygen requirement as soon as the muscle must muster 15% of its maximal strength. During static holding work, the heart rate does not attain a steady state when the force required surpasses 15% of the muscle's maximum. The greater the force applied, the steeper is the rise in heart rate. In this case also, there is a relationship to endurance. Only if a weight can be held for an unlimited duration does heart rate reach a steady state.

2.8.2 Behavior of Heart Rate after Work

After work which lies under the threshold for prolonged work, heart rate returns to its initial value within several minutes, even when this work was done over a long period. The drop in heart rate follows a negative exponential function (Fig 147). The sum of heart beats after the end of exercise which are still higher than resting heart rate usually does not exceed a total of 100. The sum of beats after exercise which still exceed the resting value is called the sum of recovery heart rates ("Erholungspulssumme" or EPS). This is an important measure for the determination of fatigue.

Heart rate after a work load which is in the range of or exceeds an apparent steady state, reacts differently. In this case, the drop in heart rate shows at least two distinct phases. There is a rapid phase, which can also be described by an exponential function, and a slow component, whereby heart rate can be elevated for 1 hour or more, depending on the duration and intensity of exercise. If the recovery heart rate sum is greater than 100, the work performed exceeds the threshold for prolonged work and is therefore fatiguing. A thorough analysis has shown that during bicycle ergometer exercise, the magnitude of the sum of recovery heart rate is proportional to the product of the amount of work performed which is greater than the threshold for prolonged work and work time. If the time required for O_2 intake, ventilation and heart rate to return to their resting values after exercise is measured, it can be seen that after work which is under the threshold for prolonged work, all variables reach their resting values at the same time. However, after exhausting work, the heart rate remains elevated 5–10 times longer than the other variables.

2.8.3 Review of the Possible Mechanisms of Heart Rate Regulation during Work

There is no doubt that the adjustment of heart rate during work can be the result of several control mechanisms. Our task is to separate these control loops from each other so that we can estimate their value during exercise. We shall first discuss a series of physiologic mechanisms which might might possibly regulate heart rate during exercise (Fig 71). The first possibility is shown by the solid lines, which indicate that the circulatory center is also innervated when the motor cortex is stimulated. The circulatory center must receive a copy of the impulses sent to the muscles if the increase in heart rate is to be proportional to motor innervation. If this mechanism actually functioned in this manner, then the question would be raised whether the form of heart rate adjustment which has been described here can be explained by this. More precise analyses have shown that the elevation of heart rate over

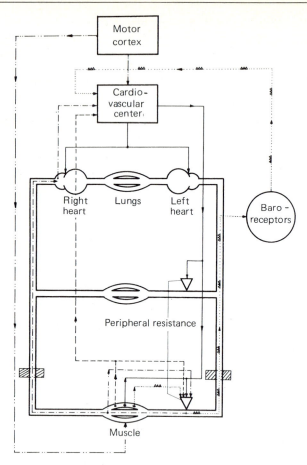

Fig 71   Series of informational possibilities which are involved with adjustment to exercise by circulatory system (see text for details).

the resting value (also called exercise heart rate) during aerobic exercise is at least proportional to the increased oxygen requirement, as well as to the work load when there is a constant rate of efficiency. The central motor system must send impulses to the circulatory center which are proportional to the work performed. The effective work performed, however, is adjusted peripherally via the muscle spindle system.

Earlier authors investigated the influence of voluntary, electric and reflex work performances on heart rate adjustment and found that exercise heart rate in all three cases was proportional to the increase in oxygen requirement. At the same time, they found that the period of adjustment was shorter for work performed voluntarily. Naturally, it was not possible to exclude the possibility that feedback from the spindles was integrated in the circulatory center and that heart rate was adjusted in this way. Therefore, if joint innervation cannot adjust heart rate in accordance with work load, then it could apparently have a role in control technology as a disturbance-input switch, i. e., although heart rate is raised at the onset of exercise, its adjustment is taken over by other control organs. Whether the rise in heart rate can be explained by a preparatory command via the influence of joint innervation or by a conditioned reflex is uncertain. In any case, the time delay caused by the slower demands of the effective control organs is resolved by this mechanism.

A second possibility is that the vasoactive metabolic products resulting from muscular work can locally diminish vascular resistance in exercising musculature (dotted line in Figure 71). As a result, there can be a pressure drop in the total arterial system, which in turn can produce a rise in heart rate via pressor receptors in the carotid sinus or aortic arch and thus an increased cardiac output. At this point, it should be mentioned that apparently under physiologic conditions this course is not followed either because the mean arterial pressure increases during physical work and does not fall, as would be expected with a proportional regulator.

The adjustment could also be the result of venous return. The blood coming in larger amounts from the dilated periphery could produce a rise in atrial pressure and initiate the so-called Bainbridge reflex (dot-dash line in Figure 11). However, pressure measurements in the vena cava and right atrium have shown that the venous and atrial pressures are not changed in healthy persons, even during heavy exercise.

There is a possibility that there are chemosensitive receptors in muscle whose activation initiates a rise in sympathetic tone and increases heart rate. This hypothesis is shown by the dashed line in Figure 71.

The methods shown here can be separated such that blood flow to a portion of the working muscles is interrupted, as is illustrated by the two cross-hatched blocks in Figure 71. The hemodynamic connection to the area of activity is abolished. If there is still a rise in heart rate during work by muscle groups whose blood supply has been interrupted, then it is unlikely under physiologic conditions that exercise heart rate is adjusted only via the pressor receptor reflexes. Such studies have been carried out on exercising calf muscles. The blood flow to these muscles was cut off by two blood pressure cuffs applied to the thighs and inflated to a pressure of 250–300 mm Hg.

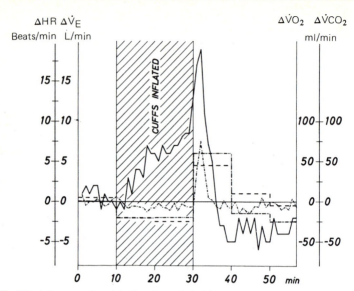

Fig 72   Influence of metabolism in occluded lower extremities on heart rate. There is a somewhat linear increase in heart rate with time after onset of occlusion. Resting values for CO_2 output and O_2 intake are reduced during the period of occlusion. The O_2 deficit of the extremities is rebreathed after ligation period ended. Both lower extremities are ligated in the shaded area of the diagram. Solid line, heart rate; dash, single dot, ventilation; dashed line, O_2 intake; dash, double dots, CO_2 output. From Stegemann, J.: Pflügers Arch. ges. Physiol. 276 : 481, 1963.

We will now consider the variations in some measurements when the cuffs were inflated under resting conditions (Fig 72). The left-hand portion of the diagram shows the initial values for O_2 intake, CO_2 production, heart rate and ventilation during the last 10 minutes before the interruption of blood flow. Inflation of the cuffs caused a rise in heart rate which continued until the tourniquet was released. The values for O_2 intake and CO_2 production were slightly lower than resting values because the metabolism of the isolated muscle group was not measured by the respiratory factors. Ventilation remained fairly constant throughout the period of compression. When the binding was removed, there was a sudden rise in heart rate, followed by a return to or under the initial value within 2–3 minutes. During this period, O_2 intake and CO_2 production increased. When the deficit contracted during application of the tourniquet at rest is compared with the amount of increased O_2 intake after the tourniquet was released, it can be seen that their values are practically identical. A similar behavior

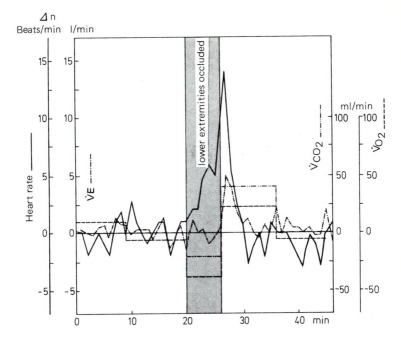

Fig 73   Increase in heart rate during occlusion of lower extremities, in which metabolism is elevated due to muscular exercise. Same symbols as in Figure 72. From Stegemann, J.: Pflügers Arch. ges. Physiol. 276 : 481, 1963.

could be observed when subjects performed mild work with calf muscles (Fig 73). The period of blood flow restricition was markedly shorter. The heart rate showed a marked rise, but during constant work this rise was linear with time. If the subjects performed varying levels of work, the larger the energy turnover of the muscle group to which the tourniquet was applied, the steeper the rise in heart rate over time.

It is possible to establish more quantitative relationships. For example, there is a linear regression between the increased volume of oxygen consumed after the release of the tourniquet and the heart rate increase during the period of ligation (Fig 74). Because the supplementary consumption of oxygen is proportional to the debt contracted during the period of restricted blood flow, the absence or surplus of a metabolic end product produced by working muscles must be responsible for the rise in heart rate. This assertion is supported by the fact that the regression is completely independent of whether the deficit or

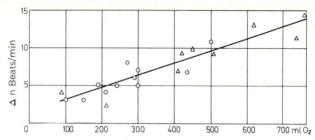

Fig 74   Regression and correlation between O_2 deficit during occlusion and increase in heart rate at rest (triangles) and during exercise (circles); R = 0.01166; r = 0.922. From Stegemann, J.: Pflügers Arch. ges. Physiol. 276 : 481, 1963.

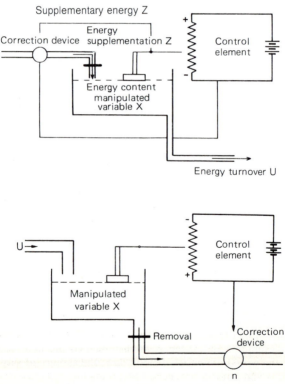

Fig 75   Functional models to explain adjustment of heart rate. For details, see text. From Stegemann, J.: Pflügers Archiv. ges. Physiol. 276 : 500, 1963.

surplus was produced by the working or resting metabolism of the musculature.

The results illustrated suggest that at least the last of the possibilities mentioned has a decisive role, namely that heart rate during exercise is adjusted by sensitive elements in the musculature. Of course, it cannot be determined at this point which substrate is responsible. On the other hand, there are several possibilities.

There are two functional models (Fig 75) which help interpret the observations. Beginning with model A, there is a concentration of an active substance X in muscle which is reduced by the energy turnover (U). If subsequent supply (Z) of this substance is blocked off, then the concentration will decrease linearly with the metabolic rate. We will now assume that the concentration of substance X is being continuously measured by chemosensitive elements and that heart rate is proportionally adjusted to it. Under these conditions and with a constant turnover rate, heart rate will increase linearly with an increasing duration of ligation; the greater the metabolic rate, the steeper will be the rise.

Model B is different from model A in that the active substance X is now the end product, which is normally carried away by the circulation. In this case, removal is hindered by the application of the tourniquet. Except for the fact that the mathematical signs are reversed, both models behave identically. Both models react in the restricted condition the same as heart rate behaves in the studies demonstrated here. How do these models respond under normal conditions, i. e., when the supply or removal can function unhindered? Assume now that the turnover rate of the muscle group is raised with unrestricted blood flow to a higher and constant value, as is physiologically found in the transition to exercise. As seen in model A, there will be a reduction in the concentration of substance, resulting in an increased supply Z. After a given time, a steady state will be attained when inflow and outflow have reached a new equilibrium. However, this new balanced flow is reached only when the concentration of substance X has decreased; then it must maintain the stimulation needed for the adjustment. It is not difficult to realize that we are talking about a proportional control loop which has the task of holding the concentration of substance X at a constant value.

The same assumptions are obviously valid for Model B, except that the mathematical signs are reversed. If the same assumptions are applied to this model, then under normal exercise conditions heart rate regulation is closely related to either oxygen debt or to carbonic acid retention by muscle; this simplification excludes the intermediary products of oxidative metabolism.

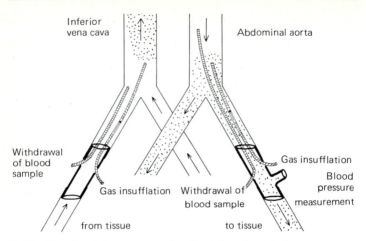

Fig 76   Experimental arrangement for stimulation of metabolic muscle receptors in animal research. From Stegemann, J., Ulmer, H.-V., Böning, D.: Pflügers Arch. ges. Physiol. 293 : 155, 1967.

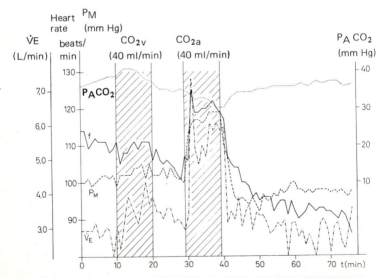

Fig 77   Effect of 10-minute CO_2 insufflation of 40 ml/min in vena cava (CO_{2v}: left shaded area) and in abdominal aorta CO_{2a}: right shaded area). f, heart rate; $\dot{V}_E$, ventilation; P_M, mean arterial blood pressure; P_ACO_2, end-tidal CO_2. From Stegemann, J., Ulmer, H.-V., Böning, D.: Pflügers Arch. ges. Physiol. 293 : 155, 1967.

2.8.4 Nature of Adequate Stimulation of Muscle Receptors

Results from animal research led initially to the hypothesis that the circulatory drive during exercise was the result of local CO_2 pressure in the musculature. If two plastic catheters are inserted into both the abdominal aorta and inferior vena cava of anesthetized dogs (Fig 76) and gaseous CO_2 is alternately bubbled into the inferior vena cava and abdominal aorta, results similar to those shown in Figure 77 will be found. When CO_2 gas is insufflated into the vein, heart rate and arterial pressure remain unchanged. End-tidal CO_2 pressure (P_ACO_2) increases slightly and ventilation rises insignificantly. Entirely different results are obtained when the same amount of CO_2 gas is bubbled into the aorta for the same length of time. In this case, there is a massive rise in heart rate and a marked increase in mean arterial pressure.

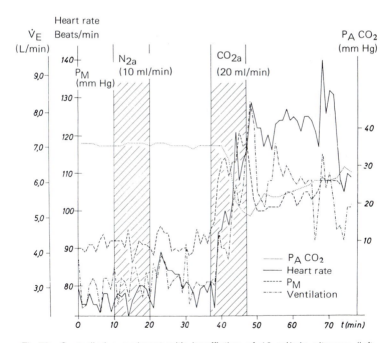

Fig 78 Controlled experiment with insufflation of 10 ml/min nitrogen (left shaded area) in abdominal aorta. There is no change in the values obtained. With further dose of CO_2 (right shaded area), there is once again a marked response, which is strongly protracted by previous administration of N_2. Same symbols as in Figure 77. From Stegemann, J., Ulmer, H.-V., Böning, D.: Pflügers Arch. ges. Physiol. 293 : 155, 1967.

Thus, the local action of CO_2 on the musculature provides a strong stimulus for the circulatory and respiratory systems.

It could be argued that such an effect is not necessarily specific to CO_2. Because gas emboli hinder blood flow, thus causing anoxia, this could have been the cause of the effects described. During the intervals indicated by crosshatching, therefore, 10 ml nitrogen per minute were insufflated into the aorta. Because nitrogen is more difficult to dissolve in blood than carbon dioxide, gas emboli in this case would be much larger. Nevertheless, there was practically no reaction (Fig 78). When CO_2 was insufflated 20 minutes later, the same respiratory and circulatory stimuli were found as described previously. This time the effect lasted much longer, probably because CO_2 could not be removed as rapidly due to partial blocking of blood flow by the previous administration of N_2 gas emboli. Perfusion of an isolated muscle group with potassium salts also provoked similar increases in heart rate. This was interpreted by other authors to mean that K^+ ions could also possibly be considered an adequate stimulus because muscle generally yields potassium during work.

When the dynamics of the transfer function (Suppl, p 305) are considered, the following is found. If work is varied in a sinusoidal fashion, then the response functions of heart rate also follow a sinusoidal pattern. Between the sinusoidal initial function and the response function, there is a variable angle of phase displacement which is frequency-dependent (Fig 79). The amplitude of the response function decreases with increasing frequency of the previous sinus function.

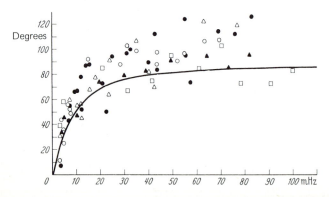

Fig 79   Dynamics of heart rate adjustment as result of sinusoidal exercise. Individual symbols represent different subjects. Ordinate denotes phase angle between sinusoidal work and heart rate as function of frequency of change in performance in mHz. Solid line is computed from model. Detailed explanations are in text. From Stegemann, J.: Pflügers Arch. ges. Physiol. 216 : 511, 1963.

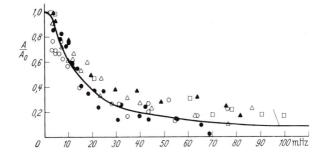

Fig 80   Dynamics of heart rate adjustment during sinusoidal exercise. Ordinate represents attenuation of sinusoidal heart rate amplitude with increasing frequency of sinusoidal exercise. Solid line is calculated from model. Detailed explanations are in text. From Stegemann, J.: Pflügers Arch. ges. Physiol. 276 : 511, 1963.

When the response amplitude obtained at a given frequency is compared to the amplitude obtained at a frequency of 0, i. e., static amplitude, then an attenuation ratio A/A_0 is obtained, in which A is the actual amplitude and A_0 is the static amplitude. In Figure 80, this ratio is plotted as a function of frequency.

If it is assumed that the signal for the adjustment of heart rate is transmitted via spinal nerves, then it is possible to investigate whether the time delay seen in the curves originates in the CNS. Studies on dogs, in which a central stump of a spinal nerve was stimulated, have shown that even the first pulse interval after the onset of stimulation was shortened. It can thus be concluded that the response delay must originate in the periphery.

The form of the model shown here corresponds to the model that was discussed in relation to oxygen debt (p 52). The behavior of heart rate over time must have something to do with the behavior of metabolic variables in the musculature over time. The simplest method would be to test whether the exponential rise in the oxygen requirement of a muscle group, as was illustrated with animal research on page 51 (Fig 31), agrees with the exponential rise in heart rate at the onset of work.

With the aid of cybernetic methods, it is simple to convert the heart rate pattern during sinusoidal work loads into the heart rate pattern during a progressive increase in work load or vice versa:

$$tg \, \psi = \omega \, \tau, \tag{39}$$

$$\frac{A}{A_0} = \frac{1}{\sqrt{1 + (\omega\tau)^2}}, \tag{40}$$

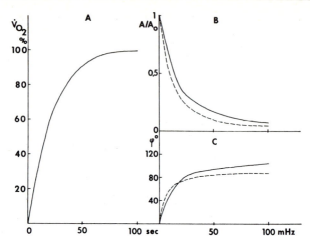

Fig 81   Comparison of relationship between oxygen consumption, heart rate and work output. A, time course of net increase in oxygen consumption in per cent of maximal steady-state value, as a response to a stepwise increase in work output. B and C, attenuation ratio (A/A$_o$) and phase angle (φ) of oxygen consumption (dashed lines), calculated from data of (A) and heart rate response (directly measured values with sinusoidal work). From Stegemann, T. and Kenner, J.: Arch. Kreislaufforsch. 64 : 185–214, 1971.

where ψ corresponds to the angle of phase difference, τ the time constant and ω the loop frequency.

On the left side of Figure 81 is the measured response function of oxygen intake during a sudden rise in work performed, as was already shown in Figure 31. For comparison, there are broken lines on the right side of the diagram which represent the response functions of the local oxygen intake during a sinusoidal variation in work load, as computed by equations 39 and 40. If this is compared to the measured curves of heart rate response to sinusoidal variations in work load, then it can be seen that both the curve for the angle of phase difference and the attenuation ratio are in close agreement relative to their pattern and their absolute values.

Attempting now to draw conclusions from these results about an adequate stimulus, one must consider which variables in the muscle follow the same time course as oxygen intake. The first thing that comes to mind is the previously mentioned study on carbon dioxide, whose pressure might represent an adequate stimulus. However, the rise in carbon dioxide pressure would occur 5–10 times more slowly because muscle has a large reserve capacity for carbon dioxide. For this reason, it is possible with some certainty to eliminate carbon dioxide pressure as an adequate stimulus. The same is also true for temperature because of the high heat capacity of muscle. Direct measurements of muscle temperature have shown that it takes about 10 minutes to reach a new steady state after the onset of submaximal exercise.

We have seen that energy-rich phosphates decrease synchronously with the rise in oxygen requirement. The best agreement is found between the change in energy-rich phosphates and the inversely related change in heart rate. There is more to suggest that heart rate adjustment and energy-rich phosphate concentration are directly correlated with one another. When an individual works slightly above his threshold for prolonged work (as was seen on p 54), there is decrease in the level of energy-rich phosphates only at the onset of exercise; this then remains constant until the glycogen reserves have been exhausted. A parallel pattern has been seen in heart rate during exercise slightly over the threshold for prolonged work and lasting many hours. At first, heart rate remains constant over several hours and then begins to rise, presumably due to the ultimate depletion of the glycogen and phosphate reserves.

During higher work loads, it is clearly seen that heart rate rises continuously with time. Figure 147 (p 244) shows the relationship between work load, threshold for prolonged work, heart rate and recovery time with bicycle ergometer exercise. It is evident that above the threshold for prolonged work, the greater the difference between the actual work load and the threshold for prolonged work, the steeper the rise in heart rate. Comparing the heart rate pattern with results shown earlier on page 55, further parallels can also be observed. The drop with time in phosphate reserve has a pattern similar to that of the rise in heart rate.

From the described relationships, the pattern of heart rate during recovery from fatiguing and nonfatiguing exercise is also clear. After nonfatiguing exercise, heart rate returns directly to its resting value. According to the model, the drop must occur at a time constant of about 25 seconds; this is actually what happens. After the completion of fatiguing exercise, the glycogen reserve is replenished. As expected, local muscle metabolism must remain elevated because ATP is required; this is based on the fact that the energy of 1 mol ATP is needed to synthesize glycogen from 1 mol glucose (p 49). This synthesis apparently follows the law of mass action and thus is essentially exponential. Finally, the energy needed for this must be taken from the aerobic metabolism. As long as this is elevated, according to the mechanisms discussed above, heart rate is also elevated. Because heart rate is involved in the recharging of both reserve systems, it is obviously a good measure of fatigue, a parameter of interest to work physiology. We will return to this subject in the chapter entitled "Physical Performance Capacity."

When discussing a correlation between a change in energy-rich phosphates in the muscle cell and the change in heart rate, it must be clearly stated that no cause-and-effect relationship has yet been found. The same is true for the adjustment of sympathetic tone, as was shown on page 102 relative to regulation of local muscle blood flow. That is, no functional receptors have been found in the muscle cell. In the interstitium there are free nerve endings which could function as receptors. In this connection, it is interesting that local infusions of a potassium solution can influence not only local peripheral resistance, but can also initiate the drive mechanisms for heart rate. The local reaction and the reaction transmitted centrally via muscle receptors show considerable parallels. As a result, the idea arises that the stimulating mechanisms in working muscle which act on the smooth vascular system and on the nervous structures are identical or at least similar. We have already shown that increasing amounts of potassium appear in the interstitium when ion pumps are inhibited by H^+ ions outside the cell. This is possibly the key to understanding the effect of CO_2 during insufflation (p 132), because a high local CO_2 pressure will elevate the number of hydrogen ions.

The other parameters mentioned (osmolality, interstitial hydrogen ion concentration and possibly even inorganic phosphate) could contribute to the effect.

2.8.5 Cardiac Output and Work

The increased sympathetic tone initiated by exercise at the level of the muscle receptors affects more than just heart rate. The rise in heart rate is only an obvious expression of a general elevation in sympathetic tone of the heart, with a recognized effect on the total myocardium. Cardiac output can increase from a value of 5 l/minute in a man at rest up to 20 l/minute during exercise by an untrained person and greater than 30 l/minute in athletes.

Using several diagrams, we will now demonstrate the adaptation of heart rate, stroke volume, peripheral resistance and AVD for oxygen (average value of 8 subjects in Figure 82) as a function of oxygen intake. Each of the values was measured 8 minutes after the onset of exercise carried out in the supine position. In the range studied, heart rate rose linearly in relation to oxygen intake (Fig 82, B). Cardiac output was also a linear function of oxygen requirement (Fig 82, A); as a result, stroke volume was essentially constant. Exact calculations show that at an oxygen intake of 1.5 l/minute, stroke volume rose only slightly from a value of 90 ml to 107 ml (Fig 82, C), and then decreased somewhat at higher levels of oxygen intake. There was a moderate rise

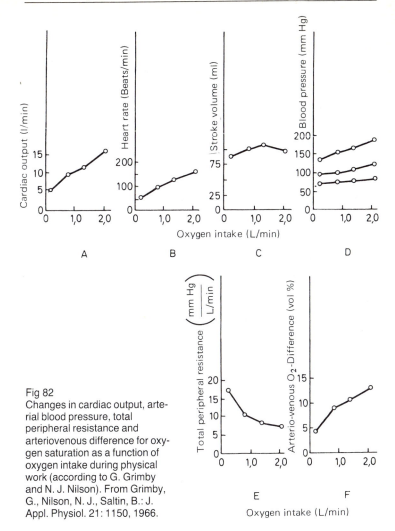

Fig 82
Changes in cardiac output, arterial blood pressure, total peripheral resistance and arteriovenous difference for oxygen saturation as a function of oxygen intake during physical work (according to G. Grimby and N. J. Nilson). From Grimby, G., Nilson, N. J., Saltin, B.: J. Appl. Physiol. 21: 1150, 1966.

in mean arterial pressure (Fig 82, D). On the other hand, systolic pressure rose to a value of 180 mm Hg at an oxygen intake of 2 l/minute, whereas diastolic pressure remained practically constant. Total peripheral resistance, calculated from these values (Fig 82, E), fell markedly during mild work loads and even more slightly during higher work loads. Finally, there was a rise in the AVD for O_2 from 4.3 vol% at rest to 12.3 vol% at heavy work.

When exercise is done in the sitting position, the values are different. Under these conditions, the resting stroke volume is smaller. It increases about 20–30% at the onset of exercise, only to fall once more at heavier work loads. At this point, it should be mentioned that cardiac output determinations in working man are plagued with certain errors in methodology; the differing results therefore are related to the method used. It is especially difficult to precisely determine stroke volume at rest because any treatment (e. g., heart catheterization) arouses the subject, which in turn stimulates the sympathetic system. As a result, measures of stroke volume at rest are often too high.

2.8.6 Blood Pressure during Physical Work

As is the case with all increases in sympathetic tone (even those caused by exercise), there are essentially two effects on the circulation: an increase in heart rate and thus cardiac output, as well as an influx of sympathetic impulses to the vasoconstrictors. Because the threshold of the vasoconstrictors in working muscles is increased for the sympathetic constrictor stimuli, the resulting vasoconstriction occurs only in

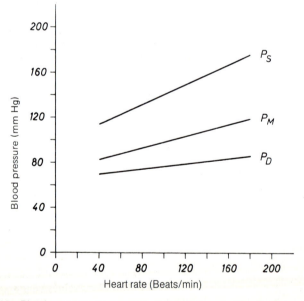

Fig 83 Blood pressure as function of physical work. On abscissa, level of heart rate is displayed as relative measure of physical stress (according to values of Holmgren).

those blood vessels which are not participating in the exercise. As a result, there is a simultaneous redistribution of blood volume primarily to the exercising regions.

During physical exercise, the blood pressure set point is adjusted to a higher level. This means that the reduction in peripheral resistance is less than the increase in cardiac output, resulting in a rise in mean blood pressure. This elevation primarily affects systolic pressure. The diastolic pressure will be primarily affected by the static components of the work being performed, i. e., it rises more steeply when the proportion of static work is higher. Figure 83 shows the behavior of systolic and diastolic pressure in a large sample of untrained subjects as a function of heart rate.

2.8.7 Sex-Related Differences in Individual Circulatory Variables

The values for individual circulatory parameters shown above were average values for untrained men. Larger series of investigations have all found that women generally have a higher heart rate at the same increase in oxygen intake. Oxygen intake and efficiency for the same types of work are the same in both sexes. Women must transport the same amount of oxygen to the working muscles with an average of 1.5 gm less hemoglobin per 100 ml; this corresponds to a smaller oxygen carrying capacity of 2 vol %. Among women, the average heart volume is smaller, so that the smaller stroke volume must be compensated by a higher heart rate. Under the presumably correct assumption that the peripheral extraction of oxygen from blood is the same in persons of both sexes at the same level of conditioning, the average stroke volume of women is about 55% of that for men.

3   Respiration

An increase in circulatory performance during exercise would be ineffective if respiration were not simultaneously amplified. After all, oxygen intake can increase from a resting value of about 250 ml/minute to 5,000 ml/minute in trained athletes. This increase corresponds to a 20-fold rise from the resting value. Respiration is concerned with supplying oxygen to the blood and eliminating carbon dioxide.

3.1 Work of Breathing and the Energy Metabolism Required

The basic condition for gas exchange, i. e., that oxygen is supplied to the lungs and that accumulated carbon dioxide is removed, results from the action of muscles to periodically expand and diminish the internal volume of the thorax. In this way, the necessary pressure difference between the inner spaces of the lung and the surrounding atmosphere is generated, producing the flow of gas. The magnitude of the work of breathing at rest requires about 1% of the basal metabolism, i. e., it is relatively unimportant. However, during exercise the metabolism for respiratory work assumes a larger fraction of the total metabolism. Thus, we must now look at the mechanics of breathing and analyze more closely the different components of respiratory work.

The lungs are situated in the thorax so that they can move easily. As a result of their elastic fibers and the surface tension of the alveoli, the lungs have the tendency to contract. They are prevented from doing so because their surface lies on the inner wall of the thorax, separated only by a capillary fluid space, the intrapleural space. Because fluid is not distensible and the intrapleural space is entirely sealed off from the atmosphere, the lungs follow the respiratory movements. Due to the elasticity of the lungs, there is a negative pressure in the intrapleural space relative to the atmosphere. The absolute magnitude of this negative pressure depends primarily on the surface tension. For this reason, the negative pressure is larger when the thorax is in the inspiratory position than in the expiratory position.

The work that the respiratory muscles must perform is composed of several components. One part is the work of deformation, which serves to distend the lungs during each inspiration. Another is the work required to alter the form of the thorax, which due to its structure

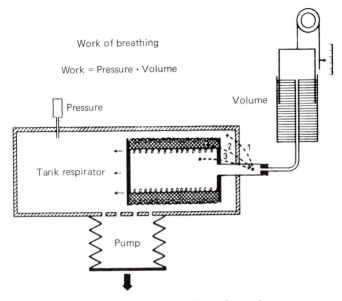

Fig 84   Definition of respiratory work (according to Comroe).

and musculature behaves like an elastic structure. Neither the lungs nor the thorax is ideally elastic, but each has plastic properties. Therefore, another type of work must be done to produce the plastic deformation. The actual work of breathing is that amount needed to produce a pressure difference in the lungs relative to the atmosphere; this enables respiratory gases to flow through the bronchial tree and trachea. Respiratory passages have a definite respiratory resistance. Respiratory work ($P \times V$) must be performed once during inspiration and again during expiration; this work includes frictional resistance in the articulations, for example. In addition, there is still another small fraction of work needed to accelerate these gases. The only way to correctly measure the total work of breathing is via artificial respiration because it is the only situation in which the actual amount of work needed to distend the thorax can be determined. Figure 84 is a diagram of how to measure respiratory work.

If the relationship between respiratory volume and the necessary associated pressure is plotted in a pressure-volume diagram (Fig 85), then the different elements of respiratory work can be labeled. At rest, the total amount of respiratory work is expended by the inspiratory muscles. Inspiratory work in Figure 85 is depicted by the shaded area. Expiratory work at rest requires no additional work by the muscles because their part is executed by the potential energy of the elastic

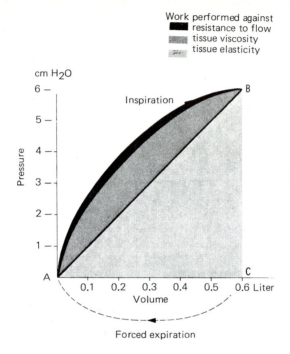

Fig 85   Pressure-volume diagram of lungs and definition of work of breathing. From Bartels, H.: Gaswechsel, In: Kurzgefasstes Lehrbuch der Physiologie, 2nd ed., ed. by Keidel, W. D. (Stuttgart: Thieme, 1970).

elements during recoil. The actual work of breathing is therefore produced by inspiratory muscles.

By amplifying the respiration during physical exercise, there are alterations in the individual work components. As a result of the general increase in sympathetic tone, the bronchial musculature relaxes, causing a reduction in resistance. The calculation of respiratory work is possible only as a rough approximation because laminar flow prevails in the bronchioles and a part of the bronchi, while in the upper airways the flow of respiratory gas is basically turbulent. Resistance to turbulent flow is greater than resistance to laminar flow. Because of this turbulence, airway resistance rises with increasing ventilation to such an extent that the real ventilatory work becomes disproportionately greater.

The rise in elastic work is essentially proportional to tidal volume. During expiration, pressure-volume work can increase so much that the force of recoil in the lungs is insufficient. This work is then assisted

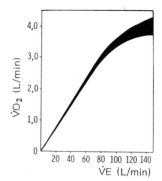

Fig 86 Relationship between ventilation and oxygen intake. Black surface is estimated proportion of oxygen intake needed for the work of breathing. From Otis, A. E.: The work of breathing, In: Handbook of Physiology, Sect. 3, vol 1, Am. Physiol. Soc. Washington D. C., 1964.

by the action of the expiratory muscles. During forced, work-related respiration, the potential energy stored during inspiration is utilized to a greater extent.

Investigations on the energy needed for breathing have found that there is an increased intake of about 0.5–1.0 ml O_2 per liter ventilation at moderate levels. The values for increased oxygen requirement were obtained during voluntary hyperventilation and cannot be applied to the same ventilatory rates as those initiated by exercise because considerably higher metabolic rates have been found with voluntary hyperventilation. Values for involuntary ventilation indicate that the efficiency of respiratory muscles is 5–20%. Figure 86 demonstrates the fraction of the total oxygen requirement associated with respiration during exercise. The upper limit of the darkened curve represents total oxygen intake and the width of the curve reflects the oxygen required by respiration. During exhausting physical exercise, at which time 120 l/minute is ventilated, about 12% of the total oxygen intake goes to the respiratory muscles.

3.2 Gas Exchange

Respiratory gases have the tendency to diffuse from regions of higher pressure to regions of lower pressure, i. e., to balance out their pressure differences. The highest oxygen pressure within the body is found in the incoming airways; the value of about 150 mm Hg is practically the same as atmospheric pressure. The lowest pressure is found in the mitochondria. This pressure gradient is the reason oxygen passes from the airways to the tissue cells. Inverse relationships are present for the end products of metabolism. For example, carbon dioxide produced in the body's cells is at a much higher pressure than that in the atmosphere (0.03 vol %).

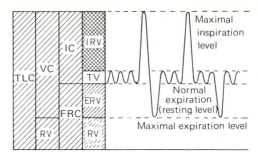

Fig 87   Nomenclature of different lung volumes and capacities. TLC, total lung capacity: largest possible amount of air contained in lungs after maximal inspiration; VC, vital capacity: volume between deepest inspiration and expiration; RV, residual volume: volume of air remaining in lungs after deepest expiration; IC, inspiratory capacity: volume between normal resting expiratory level and maximal inspiration; FRC, functional residual capacity: volume of air remaining in lungs after normal expiration (resting level); IRV, inspiratory reserve volume: volume between that after normal inspiration and that after maximal inspiration; TV, tidal volume: volume of air moved during normal respiration; ERV, expiratory reserve volume: volume between that after normal expiration and that after maximal expiration. From Wiemers, K., Kern, K., Günther, M., Burchardi, H.: Postoperative Frühkomplikationen, 2nd ed. (Stuttgart: Thieme, 1969).

Diffusion over the great distances found in the body would not be sufficient for supplying the tissues if there were not a series of mechanisms which actively amplify this pressure gradient and adapt themselves to the existing needs of the periphery. An important part of this active mechanism is the periodic partial renewal of the volume of air in the lungs.

Figure 87 shows the magnitude and nomenclature of the individual volumes and capacities which are important for the comprehension of gas exchange. At rest, there are about 3–3.5 l air in the lungs (middle breathing position). Tidal volume is that volume of air which passes into and out of the lungs with each breath and is approximately 0.5 l. After a maximal expiration, an individual can exhale about 2 l starting from the midposition (expiratory reserve volume). Nevertheless, there is always still another 1.2 l remaining in the lungs (residual volume) which cannot be voluntarily exhaled. After the deepest inspiration starting from the midposition, an additional 2 l can be inhaled (inspiratory reserve volume). The maximal amount of air which can be voluntarily ventilated in one breath is called vital capacity. Adding the residual volume to this gives the value for total lung capacity. The absolute magnitude of these various volumes is related not only to body height and age, but also to the level of physical training.

During physical exercise, there is little change in the mid-breathing position, while the tidal volume uniformly increases at the expense of the expiratory and inspiratory reserve volume. Based on the fact that a portion of the volume in the inner lung space is renewed with each breath, there is also a cyclic variation in the concentration and the pressure of the respiratory gases. Carbon dioxide coming from the blood causes a further increases in alveolar P_{CO_2} in the first phase of inspiration because only air from the dead space flows into the lungs and it has the same concentration as that found in alveolar air at the end of the previous expiratory phase. Dead space is the portion containing the air which does not participate in gas exchange. It is the space encompassed by the upper airways, the oral cavity and the bronchi.

As soon as fresh air flows into the lungs, there is a sudden drop in alveolar CO_2 concentration, reaching a minimum at the end of inspiration. During expiration, the concentration increases once more.

The variations in alveolar concentration depend on CO_2 production, respiratory frequency and tidal volume; all three of these variables change during physical work. The fact that with increasing CO_2 production the air cushion decreases because of the increased use of the inspiratory reserve volume means that the cyclic variations in alveolar concentration are larger during exercise. These variations originate in the pulmonary veins but are spatially and temporally integrated by the intracardiac volume in such a way that the mean arterial P_{CO_2} is less than the maximal alveolar CO_2 pressure, but more than the minimal value. Real pressure gradients caused by resistance to diffusion in the alveoli do not have any practical role due to the high solubility of CO_2.

3.3 Respiratory Dead Space and the Concept of Alveolar Ventilation

It must be clearly understood that an increased ventilation does not necessarily lead to an elevated elimination of CO_2. Shallow ventilation, such as seen when a dog pants, can be large but it does not necessarily result in an exchange of gases because the air only moves in and out of the dead space and is not renewed.

The composition of air in the dead space is not the same as inspired air and is in no way separated from the effective alveolar space by a sharply defined border at the entrance of the alveoli. More often, there is a continual intermixing of the two gaseous phases. This is basically produced by two factors: a diffusion related to the pressure gradients of O_2 and CO_2 and the flow relationships present in the airways. With laminar flow, for example, the flow in the central area is faster than

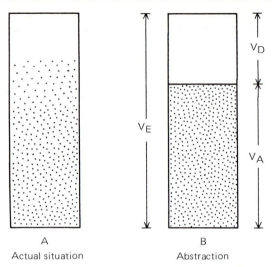

A
Actual situation

B
Abstraction

Fig 88   Schematic explanation of terms used in alveolar ventilation (according to Rossier, Bühlmann, Wiesinger).

that at the edge. The diffusion of CO_2 from the alveolar space toward the dead space (and of O_2 in the opposite direction) is affected so that a portion of the air from the anatomical dead space can still be made available for gas exchange. In the same sense, laminar flow accomplishes this during inspiration because fresh air is able to reach the alveoli via the faster central flow before the air in the dead space has been completely reinspired. In the opposite way during expiration, alveolar air can already be exhaled before the dead space is completely washed out.

This is the only way to explain why the alveoli can also be ventilated with shallow breathing, even when the tidal volume is smaller than the volume of the anatomical dead space. The extent to which alveolar air is renewed by a breath does not depend on the magnitude of the anatomical dead space but on the size of the functional dead space. Functional dead space does not imply any space in the geometric sense of the word, only a theoretical abstraction, as illustrated in Figure 88. Immediately before expiration, the ratios correspond to those in Figure 88, A, i. e., the alveolar air is transformed into environmental air, such that the O_2 content continually increases and the CO_2 content continually decreases, going from the interior toward the exterior. The specifics of this transition cannot be quantitatively determined with precision. Accordingly, the tendency is to visualize a part of the lungs filled with only alveolar air and the other part filled with only environmental air (Fig 88, B). On the basis of the two columns, the same gas concentration prevails, i. e., area × intensity is identical in both columns. This diagram also explains the calculation of physiologic dead space according to the formula of Bohr, who assumed a constant dead space of 140 ml and then calculated the gas concentration of alveolar air by the formula:

$$V_D C_i = V_T C_E - (V_T - V_D) C_A. \qquad (41)$$

In this formula, V_T is tidal volume, V_D is dead space volume and C_A, C_E and C_i are the average gas concentrations in alveolar, expiratory and inspiratory air, respectively. The Bohr formula means that the quantity of expired gases ($V_T \times C_E$) can be divided into two parts. One part ($V_D \times C_i$), corresponds to dead space times the concentration in inspired air, and the second part ($V_T - V_D) \times C_A$, comes from the alveolar space and produces the concentration of gases in the alveoli. Three of the 5 variables in the formula (V_T, C_i and C_E) are easy to measure. To determine dead space V_D, the alveolar concentration C_A must be known.

Assuming that there is no CO_2 in inspired air, the equation for this gas is simplified:

$$V_T C_E = (V_T - V_D) C_A \quad \text{or} \qquad (42)$$

$$V_T C_E = V_A C_A. \qquad (43)$$

Solving the equation for V_A yields:

$$V_A = \frac{V_T C_E}{C_A} \qquad (44)$$

or when it is multiplied by the respiratory frequency and the alveolar concentration is converted to alveolar pressure:

$$\dot{V}_A = \frac{\dot{V}_{CO_2} \dfrac{760 \cdot (273 + t)}{273}}{P_{A_{CO_2}}}. \qquad (45)$$

If instead of using the imprecisely determined alveolar CO_2 pressure ($P_{A_{CO_2}}$), the average arterial CO_2 pressure ($P_{a_{CO_2}}$) is inserted into the equation, then at a body temperature (t) of 37 C the following formula is obtained:

$$\dot{V}_A = \frac{\dot{V}_{CO_2} \cdot 863}{P_{A_{CO_2}}}, \qquad (46)$$

where $\dot{V}_A$ is measured under BTPS conditions (BTPS is body temperature and pressure, saturated: 37 C, ambient atmospheric pressure and saturated with water vapor) and $\dot{V}_{CO_2}$ is measured under STPD conditions (STPD is standard temperature and pressure, dry: 760 mm Hg, 0 C, dry).

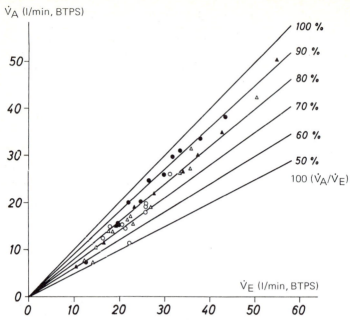

Fig 89   Alveolar efficiency as function of work performed on bicycle ergometer. From Stegemann, J. and Heinrich, K. W.: Studien über den respiratorischen Totraum bei körperlicher Arbeit und bei künstlicher Beatmung (Cologne: Westdeutscher Verlag, 1967).

The product of the effective space V_A and respiratory frequency is called alveolar ventilation ($\dot{V}_A$). Under the conditions of a constant production of CO_2 alveolar ventilation influences alveolar and arterial CO_2 pressures. The larger the alveolar ventilation at the same total ventilation ($\dot{V}_E$), the more economic is the breathing because only alveolar ventilation is of importance for gas exchange. The panting dog breathes with a large total ventilation but with a low relative alveolar ventilation. The formula, $\dot{V}_A/\dot{V}_E$ gives information about the economy of breathing and is therefore called alveolar efficiency. Under resting conditions, the V_E of a man weighing 70 kg is about 7 l/minute; of this, about 2 l/minute is dead space ventilation and the other 5 l/minute is the $\dot{V}_A$. From this, it can be seen that alveolar efficiency is about 70%. Figure 89 shows the effects of increased ventilation brought on by exercise on alveolar efficiency. With increasing work loads, alveolar efficiency rises. Breathing economy with increasing work is therefore improved not only in regard to respiratory work (p 144), but also in terms of gas exchange.

Table 12   Composition of Fresh Air, Expired Air and Alveolar Air

	Fresh Air (vol%)	Expired Air (vol%)	Alveolar Air (vol%)
Oxygen	20.95	16–17	14–15
Nitrogen and inert gases	79.02	79–81	79–81
Carbon dioxide	0.03	3–4	5–6

Even when there is an almost complete pressure equalization for CO_2 between blood and alveolar air, there can be differences between alveolar CO_2 pressure and CO_2 pressure in the left side of the heart or peripheral arteries. These arterial-alveolar pressure gradients can be caused by the fact that venous blood is mixed with arterial blood. For example, a small arteriovenous short circuit can result from the venae cordis minimae (thebesian veins of the heart) which empty into the atrium or from anastomoses with bronchial veins. More important for the question of the functional dead space, however, is the relationship between ventilation and blood flow in the individual alveoli, the "ventilation-perfusion ratio." A perfused alveolus with little or no ventilation results in an arterial CO_2 pressure which is greater than that in the alveoli because these alveoli supply pulmonary venous blood which has not released its CO_2. A ventilated alveolus with no blood supply means that the alveolar P_{CO_2} is less than the arterial P_{CO_2} because such an alveolus becomes part of the functional dead space.

For oxygen, the alveolocapillary membrane represents a real resistance to diffusion. The pressure gradient between arterial and alveolar O_2 pressures is 6–8 mm Hg P_{O_2}, taking the short circuits into consideration. This difference remains essentially unchanged, even during strenuous physical exercise by healthy persons, because the larger cardiac output to the lungs is compensated by the larger active exchange surface.

Table 12 gives the normal values of gas concentration in atmospheric air, in expired air and in alveolar air under resting conditions.

3.4 Relationship between CO₂ Dissociation and Buffering

To better understand the total regulatory process step by step, the dissociation of CO_2 (which has alredy been introduced on p 96) should be briefly summarized. If blood is set in equilibrium with CO_2, the greater the pressure in the equilibrating gas, the more CO_2 will be physically dissolved in blood. There is a proportional relationship between the amount of dissolved gas and the pressure, as seen in the equation:

$$[CO_2] \rightleftarrows \alpha P_{CO_2} \qquad (47)$$

The proportionality constant α depends on temperature and is called the Bunsen solubility coefficient. For plasma at a temperature of 37 C, α equals 0.53 ml CO_2/ml $H_2O \times$ atmosphere P_{CO_2}.
CO_2 is hydrated only from the fluid component of blood:

$$CO_2 + H_2O \leftrightarrows H_2CO_3. \qquad (48)$$

This reaction is accelerated by the enzyme carboanhydratase found in erythrocytes. Carbonic acid now partially dissociates into H^+ and HCO_3^- ions. The constant K' reflects the ratio of dissociated to nondissociated portions:

$$K' = \frac{[H^+] \ [HCO_3^-]}{[H_2CO_3]}. \qquad (49)$$

Solving for H^+ yields the equation named after Henderson and Hasselbalch:

$$[H^+] = K' \frac{[H_2CO_3]}{[HCO_3^-]}. \qquad (50)$$

If the negative decadic logarithm is formed from H^+ and K', then the equation is:

$$pH = pK' + \log \frac{[HCO_3^-]}{[H_2CO_3]}. \qquad (51)$$

pK' has the value of 6.10. As a result, with the physiologic ratio of HCO_3^- : H_2CO_3 equal to 20 : 1, blood pH is 7.4 (log 20 = 1.30). The ratio of H_2CO_3 to HCO_3^- in plasma is essentially determined by the available alkali ions (Na^+ and K^+). At rest, two acids compete with each other: the weak "protein acid" and carbonic acid. When the carbon dioxide pressure is 0, then total blood protein is present as salts of potassium and sodium. With increasing pressure of carbonic acid, the K-proteinate in the erythrocytes reacts with the assistance of carboanhydratase in the following manner:

$$K\text{-proteinate} + H_2CO_3 = H\text{-proteinate} + KHCO_3. \qquad (52)$$

With a change in osmotic pressure, there is a flow of water into the erythrocytes. As a result of the disturbed concentration gradients, HCO_3^- ions diffuse into plasma and Cl^- ions diffuse into the erythrocytes. The erythrocytes are therefore necessary for the formation of sodium bicarbonate.
Drawing a diagram such as Figure 90 should clarify these facts. In the lower portion of the diagrams, the physically dissolved fraction of carbon dioxide is represented as a function of CO_2 pressure. Because

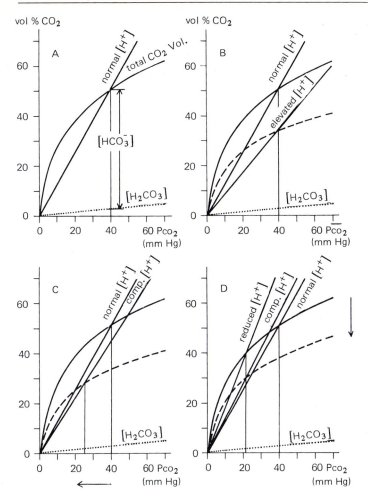

Fig 90   A, normal dissociation curve; B, noncompensated metabolic acidosis; C, compensated metabolic acidosis; D, compensated respiratory alkalosis.

the reaction between H$_2$CO$_3$ and P$_{CO_2}$ is proportional, the function is a straight line. The upper curve represents the total content of carbon dioxide. Between the two curves is the fraction of carbon dioxide which is chemically bound.

In addition to this, an isohydre is drawn in each diagram. The term isohydre is understood to be a line of equal hydrogen ion concentration. The $[H^+]$ is determined by the ratio of $[H_2CO_3]$ to $[HCO_3^-]$. A line is inserted which runs from the origin to definite points on the curve (shown here at 40 mm Hg P_{CO_2}), where the ratio and thus the $[H^+]$ are both equal.

3.5 Influence of Lactic Acid on Acid-Base Balance In Vitro and In Vivo

Under anaerobic metabolic conditions, there is a production of lactic acid. From the viewpoint of exercise physiology, we must investigate the effect of this acid on acid-base balance. Lactic acid has a higher dissociation constant than carbonic acid. As a result, depending on the concentration, it binds a smaller or larger fraction of the available alkali ions. As a consequence, less bicarbonate is formed at the same CO_2 pressure. The CO_2 dissociation curve is thus flatter (dashed curve of Figure 90 B). This condition is called metabolic acidosis. The isohydre shows that the $[H^+]$ is elevated. This is as far as we can determine the ratio *in vitro*.

In vivo, an active regulatory mechanism is introduced because the elevated $[H^+]$ is a form of respiratory stimulus (p 159) which produces increased ventilation. With a constant production of CO_2, this increased ventilation necessarily causes a drop in the P_{CO_2} of arterial blood. This fall is represented by an arrow in Figure 90, C. If another isohydre is inserted at the newly-established ratio of salt to acid at a P_{CO_2} of 25 mm Hg, then it can be seen that the hydrogen ion concentration is largely compensated, despite a flatter dissociation curve and even with a markedly higher ventilation. This condition is called compensated metabolic acidosis. The appearance of fixed acids is thus compensated by the elimination of volatile acids.

As we will see in a later section, a stay at high altitude produces hyperventilation due to the low O_2 pressure. A person cannot adapt to this hyperventilation over a prolonged period. As a consequence of the ascent, there is a lower than normal P_{CO_2} in arterial blood. This condition is schematically shown in Figure 90, D. Because no lactic acid is present at rest, the pattern of the CO_2 dissociation curve is also not changed. If another isohydre is drawn with the normal CO_2 dissociation curve, it becomes clear that the $[H^+]$ is lowered by the hyperventilation (isohydre: reduced $[H^+]$). If this condition called respiratory alkalosis is maintained for several hours, the kidney begins to eliminate increased amounts of alkaline ions. As a result, the CO_2 dissociation curve is flatter (*arrow*) and the hydrogen ion concentration is again largely compensated despite the elevated ventilation. This condition is referred to as compensated respiratory alkalosis. The buffering capacity against fixed acids is reduced during high work loads at altitude.

3.6 Estimation of Acid-Base Status from Standard Bicarbonate

The pattern of the CO_2 dissociation curve is a measure of the acid-base status of the blood. To eliminate having to determine the total curve each time, look at a definite point on the dissociation curve. Because the binding capacity of blood for CO_2 is related to the hemoglobin level and to its saturation with O_2 and because HbO_2 has a higher dissociation constant than reduced hemoglobin, the determination is made of the CO_2 bound in plasma whose blood was previously 100% saturated with O_2. In addition, the concentration of bound CO_2 is dependent on temperature (reaction equilibrium) and on P_{CO_2}. As a standard, it has been determined that blood at 37 C and a P_{CO_2} of 40 mm Hg is equilibrated even before the plasma has been obtained by centrifugation.

Standard bicarbonate is the concentration of CO_2 bound in plasma whose blood was equilibrated with a mixture of O_2 and CO_2 at a P_{CO_2} of 40 mm Hg and at 37 C. Figure 91 should clarify the term standard bicarbonate. A reduction in standard bicarbonate always means an acidosis and an increase means alkalosis, whereby alkalosis and acidosis are understood to be either metabolic or a compensation for respiratory acidosis or alkalosis. Noncompensated respiratory alkalosis or acidosis cannot be determined from standard bicarbonate because by definition this value is determined at a P_{CO_2} of 40 mm Hg.

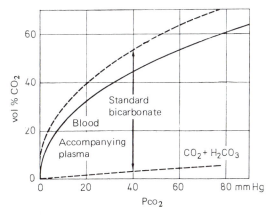

Fig 91   Definition of standard bicarbonate (according to Schneider).

3.6.1 Practical Determination of Acid-Base Balance

Now that we have learned the basics of acid-base balance, we would like to discuss its practical determination. This is particularly applicable because acid-base balance has become more and more important for the physiology of exercise and sports due to the simplicity of its determination. To understand the practical method of determination, begin with equation 51:

$$pH = pK' + \log \frac{[HCO_3^-]}{[H_2CO_3]} \tag{53}$$

According to equations 47 and 48, $[H_2CO_3] = \alpha P_{CO_2}$; α represents the Bunsen solubility coefficient in the equation. The H_2CO_3 was previously expressed in vol %. Today, it is no longer expressed in vol %, but in mEq/l. One mol CO_2 assumes a volume of 22.4 l. To convert vol % H_2CO_3 into mEq/l, the numerical value of vol % must be divided by 2.24.

The following formula is valid:

$$[H_2CO_3] \text{ mEq/l} = \frac{\alpha P_{CO_2}}{2.24}. \tag{54}$$

The term $\alpha/2.24$ is called the modified Bunsen solubility coefficient α_M. In plasma, it has a value of 0.03.

Accordingly: $[H_2CO_3] \text{ mEq/l} = \alpha_M P_{CO_2}.$ (55)

If this equation is inserted into equation 53, we obtain:

$$pH = -\log \alpha_M P_{CO_2} + pK' + \log [HCO_3^-]. \tag{56}$$

Using a diagram with the pH value on the abscissa and $\log P_{CO_2}$ on the ordinate, this relationship is linear at a constant concentration of HCO_3^- because the equation follows the form $y = -ax + b$.

If semilogarithmic paper is used, it is possible to record isobicarbonate lines (Fig 92). Investigations have shown that in blood as well, the relationship between pH and $\log P_{CO_2}$ (at least within normal physiologic limits) can be described by a straight line. Because of the inconstancy of bicarbonate concentration, this line does not run parallel to the isobicarbonate lines, but cuts across them.

A straight line can be defined by two points. To determine the unknown P_{CO_2} in the blood, three blood samples are taken from subjects in the shortest possible interval using heparinized glass capillary tubes. From the first blood sample, pH is determined. The other two blood samples are equilibrated at two known CO_2 pressures (CO_2-O_2 mixtures) and their pH value is also determined. From the known P_{CO_2} values and the corresponding pH values, the line can be constructed and the unknown P_{CO_2} value sought from its actual pH (Fig 93). In this way, the pressure of the volatile acid CO_2 can be precisely determined.

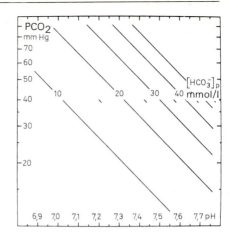

Fig 92
Relations among bicarbonate content, logarithm of CO_2 pressure and pH value of solutions containing bicarbonate.

According to equation 56, the actual bicarbonate can be calculated

$$\log [HCO_3^-] = pH - pK' + \log \alpha_M P_{CO_2} \qquad (57)$$

or even the standard bicarbonate, if the P_{CO_2} of 40 mm Hg is used.

It is simpler to proceed with a graphical solution in which the standard bicarbonate is read directly from a nomogram using the point of intersection of the pH-log P_{CO_2} lines with the isobaric line ($P_{CO_2} = 40$ mm Hg). Further details of this technique can be found in the publication of Siggaard-Andersen.

Today, there are commercial instruments using this procedure with which the acid-base status (P_{CO_2} and standard bicarbonate) can be easily determined within about 10 minutes using a blood sample of 100 µl.

3.7 Control of Breathing

The system for respiratory control is complicated and confused, in which individual influences are modified by each other. For this reason, it is difficult to define the manipulated variable, i. e., the variable which should be controlled, without taking into consideration the intermingling. Since the beginning of this century, there have been bitter feuds between individual schools of physiologic thought about whether the CO_2 pressure in arterial blood or the hydrogen ion concentration was held constant. According to the knowledge available today, it is still not so simple to decide because each of these variables, along with other forms of reference input, can affect the set point of both variables. As explained in the supplement (p 303), the

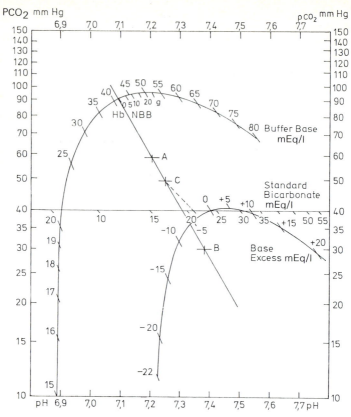

Fig 93 Siggaard-Andersen diagram for determination of CO_2 pressure in blood with the Astrup method. Copyright by Radiometer AS, Copenhagen.

sum of information presented to the regulator is the decisive factor for its output. Figure 94 shows a simplified control diagram. At the entrance to the regulator are not only the output error (x) and the reference input (y), but also several other inputs which receive negative feedback from the controlled process. In terms of an analysis of the controlled response, one of these variables can be considered the controlled variable, whereas the others are regarded as reference input variables. Which variable ultimately is considered to be the actual controlled variable is of little consequence. Because the CO_2 effect was the first to be discovered, from a historical point of view it is usually considered to be the controlled variable, whereas the other inputs act as reference input variables which modify the set point of P_{CO_2}.

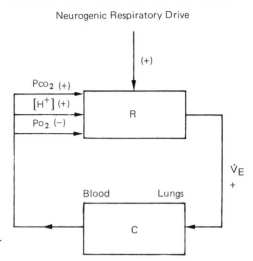

Fig 94
Block diagram of respiration. Regulator R (respiratory center) is affected by P_{CO_2}, H ions and reduced PO_2 to stimulate breathing, resulting in elevation of total ventilation ($\dot{V}_E$). This, in turn, produces increased blowing-off of CO_2 and thus a reduction in hydrogen ion concentration and a rise in PO_2. Additional neurogenic respiratory drives stimulate respiratory center during exercise. More detailed explanations are in text.

3.7.1 Control of P_{CO_2}

It is relatively easy to determine the effect of CO_2 on ventilation by rebreathing from a sack initially filled with pure O_2. The concentration of CO_2 in the sack slowly increases due to the metabolically produced CO_2. There is a gradual increase in ventilation, reaching about 60 l/minute at 6% CO_2. Above this value, ventilation begins to drop once more.

The relationship between ventilation and P_{CO_2}, as it occurs within the controlled system, has been demonstrated in the discussion of alveolar concentration. Nevertheless, the schematic diagram in Figure 95 gives an approximation of the relationships. The P_{CO_2} in arterial blood (symbolized by the water level) depends on two basic influences: the inflow of CO_2 into the blood and its release from the blood via respiration. The inflow is primarily the result of metabolism, but can also be produced by the breakdown of bicarbonate, e. g., it appears

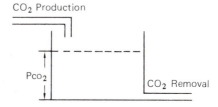

Fig 95
Model to assist understanding of relationship among CO_2 production, CO_2 removal and CO_2 pressure.

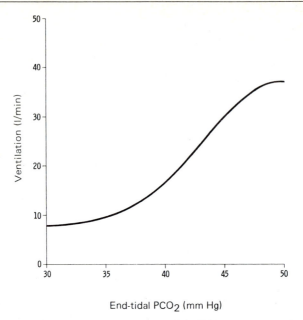

End-tidal PCO$_2$ (mm Hg)

Fig 96   Ventilation as function of end-tidal PCO$_2$ (CO$_2$ response curve).

when acids with higher dissociation constants (for example, lactic acid) are present in the blood. The release of CO$_2$ is varied because the alveolar air which contains CO$_2$ is frequently and extensively renewed with each breath.

Carbon dioxide cannot be released into the atmosphere during the rebreathing experiment. Therefore, it can be used to functionally interrupt the control loop so that the influence of P_{CO_2} on ventilation via the regulator can be studied. The result of such an experiment is shown in Figure 96. The "CO$_2$ response curve" obtained depicts the static curve of the regulator (R_1 in Fig 97). From this curve, it can be concluded that there is a nonlinear relationship between the two variables. In the classic description, there is a threshold range of about 35–40 mm Hg P_{CO_2}. The curve then has a more or less linear portion, reaching a saturation range at about 50–60 mm Hg. A more exact mathematical-statistical analysis of the form shows that it is a section of a Gaussian bell-shaped curve, possibly based on a receptor system which exhibits differing response thresholds and a limited measuring range within each of the individual receptors. The measuring sensors are located in the medulla oblongata at the base of the 4th ventricle.

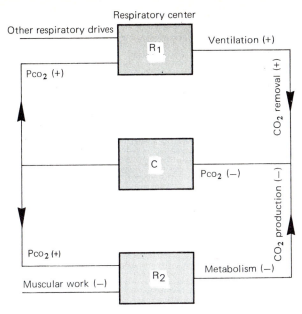

Fig 97   Schematic diagram of negative-feedback control system for maintaining constant PCO_2.

They are probably not stimulated directly by CO_2 but by the hydrogen ions which appear in the receptor cells and in the cerebrospinal fluid.
Another regulatory process (R_2 in Fig 97) has a role, at least under resting conditions. With hypercapnia, the total metabolism is reduced, whereas it is increased with hypocapnia. Whether this effect is due to a metabolic center or whether the end product CO_2 inhibits the normal metabolic pattern via the law of mass action cannot be conclusively ascertained at this time.

3.7.2 Effect of Oxygen Deficiency and pH Reduction on the CO_2 Response Curve

An effective O_2 deficiency in this instance refers to a reduced O_2 pressure of inspired air, resulting in a lower oxygen saturation of the blood. In the section on altitude acclimatization (see p 171), the relationship between O_2 pressure and altitude is discussed.
The measuring sensor for O_2 pressure is located in the carotid bodies found near the bifurcation of the common carotid artery into the internal and external carotid arteries. Other sensors are located in the

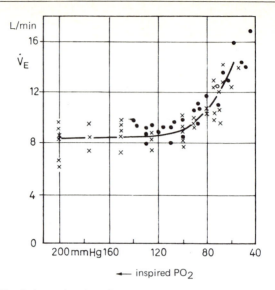

Fig 98   Ventilation as function of oxygen pressure in inspired air (according to Hartmann).

aortic body in the area of the aortic arch. The effect of reduced O_2 pressure on ventilation can be seen in Figure 98.

The CO_2 response curve exhibits a somewhat parallel shift to the left when there is a drop in O_2 pressure. This means that the effect of P_{O_2} is algebraically added to that of P_{CO_2}, i. e., the sum of the reference input (in this case, a drop in P_{O_2}) and the output error (deviation in P_{CO_2}) is effective, at least within the average, accessible range studied.

An increased hydrogen ion concentration in arterial blood has a similar effect. This effect is also measured primarily by the receptors in the aortic and carotid bodies.

3.7.3 Influence of Work on the CO_2 Response Curve

If breathing response curves are recorded at different levels of aerobic work, then it can be seen that similar to other effects, they also exhibit a parallel shift to the left (Fig 99). This important fact demonstrates that the stimulatory effect of exercise on breathing is not produced by the increased accumulation of CO_2 during exercise, but that there is a completely independent and additive effective stimulation, in the sense of an adjustment of the set point via the influence of the reference inputs.

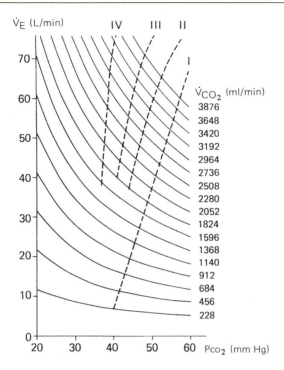

$\dot{V}_E$ (L/min)

IV   III   II

I

$\dot{V}_{CO_2}$ (ml/min)

3876
3648
3420
3192
2964
2736
2508
2280
2052
1824
1596
1368
1140
912
684
456
228

P_{CO_2} (mm Hg)

Fig 99   Respiratory response curves: Rest I; different levels of work II < III < IV. Abscissa is arterial CO_2 pressure and ordinate is ventilation; variable is CO_2 given off. From Stegemann, J.: Leistungsphysiologie, In.: Kurzgefasstes Lehrbuch der Physiologie, 4th ed., ed. by Keidel, W. D. (Stuttgart: Thieme, 1975).

Although specialists do not completely agree, it is now possible to conclude that the respiratory stimulus of exercise originates in the working muscle and is probably controlled via the same receptor system and the same adequate stimulus as heart rate (p 125). The relationships with respiration are not so clear as those regarding the steering effect of exercise on heart rate, however, because there are difficulties in interpretation due to the interference of different respiratory stimuli, when all of the controlling parameters are not measured.

To become acquainted with these difficulties, consider again Figures 72 and 73. The results displayed are based on the following methodology. In the case of the studies reported in Figure 72, heart rate and ventilation were recorded at rest, followed by the interruption of blood flow to a portion of the total musculature; this compression was later

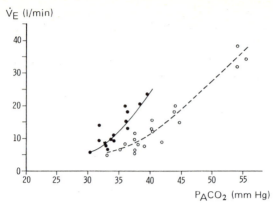

Fig 100   Ventilation as a function of alveolar PCO_2 under normal conditions (right curve) and after ligation of both legs (left curve).

released. In Figure 73, the procedure was similar, the difference being that the ligated muscles were now exercised. During the work period, there was an O_2 deficit which was proportional to the rise in heart rate during this period of compression; ventilation was not affected. The hypotheses on how this deficit influenced the circulatory center via intermediate connections were discussed on page 138. From a systematic theoretical viewpoint, it would be premature to conclude that the local constriction of a portion of the musculature has an effect on heart rate but not on ventilation because information from muscle could reach both systems but have different effects. For example, the influence could be effective in one system but not in the other because there is compensation by another negative influence. One thing certain is that during constriction of muscle, information is transmitted to the circulatory center which produces a rise in heart rate. The same effect is initiated with the ventilation. However, because the CO_2 normally liberated by muscle remains there due to the constriction, the central CO_2 drive is less (refer to Figure 94), i. e., the flow of CO_2 into the arterial system is less, resulting in a lower arterial P_{CO_2}.

To demonstrate this, Figure 100 contains response curves taken under normal conditions and after 20 minutes constriction of the lower extremities. The CO_2 response is shifted to the left after constriction. Therefore, an additional stimulus from the ligated muscles was effective.

Just as clear is the effect during exercise on a bicycle ergometer where blood flow was out off for a short period by thigh cuffs. Exercise ventilation remained constant and P_{CO_2} fell sharply. If P_{CO_2} was artificially held constant during the constriction period using a mixture of

CO_2 in the inspired air, then ventilation increased to higher levels. In this case as well, summation of both stimuli is clearly observed.

3.7.4 Behavior of Respiratory Parameters during Mild and Severe Physical Work

To understand the interplay of factors which regulate respiration, we will now consider the results of an investigation in which continuous

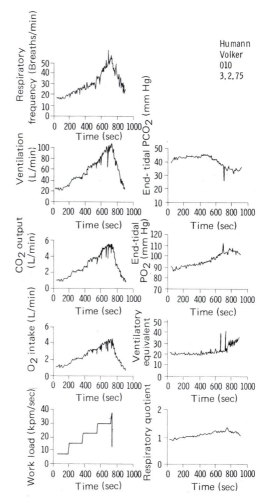

Humann
Volker
010
3.2.75

Fig 101
Recording of experiment measuring "maximal oxygen intake" during progressively increasing work loads.

measurements of ventilation and its accompanying parameters were obtained using breath-by-breath analysis (Suppl, p 315 and Fig 101). The results are from a moderately-trained student of physical education. On the lower left is work load in kpm/second on the bicycle ergometer at a pedaling frequency of 75 rpm. Every 3 minutes, work load was increased by 7 kpm/second until the subject was exhausted. Above this graph is one depicting oxygen intake, which rose to a maximal value of about 4 l/minute as a result of the increasing work load. The production of CO_2 is less at the lower work loads and greater at the higher work loads than the corresponding oxygen intake; this is also seen in the graph for R ($\dot{V}CO_2/\dot{V}O_2$). The next graph depicts ventilation, which showed a disproportionate rise after a work load of 14 kpm/second. Ventilation reached a maximal value of about 100 l/minute. The corresponding pattern of respiratory frequency can also be seen.

End-tidal O_2 increases when the threshold for prolonged work (p 242) is surpassed; in this case, it occurred at about 12 kpm/second. The reason for this is a relative hyperventilation, i. e., the respiratory drive and thus the ventilation are greater than the amount needed for the oxygen consumed.

This finding is also clearly shown in the recording of the ventilatory equivalent, i. e., ventilation (l/minute) divided by the amount of O_2 consumed (l/minute). This equivalent is used by many working in sports medicine as a measure of the economy of breathing.

The fact that P_{CO_2} drops also shows that there are additional active respiratory stimuli above the threshold for prolonged work which produce an adjustment of the set point for P_{CO_2}. This stimulation is derived from two factors: H^+ ions which appear in the blood as a result of the partial anaerobic energy supply and the disproportionate activation of muscle receptors as a result of insufficient local muscle blood flow.

3.8 Adjustment of the Breathing Pattern

Recent results suggest that a series of mechanisms are involved in the adjustment of breathing patterns, i. e., the adaptation of frequency and depth of breathing. However, it is difficult to precisely estimate their value for man because there are apparently large differences among species.

Most likely, there is a rhythmic development of stimuli within the respiratory center itself, located in the medulla oblongata. An auto-rhythmicity of inspiratory and expiratory neurons has been detected here; this is a kind of respiratory pacemaker for the basic rhythm, which can be modified by a number of afferent impulses.

In animal experiments at least, there seems to be a reflex arc, whose stretch sensors are located in the bronchioles. When they are stretched along with the lungs, inspiration is inhibited. Because exhalation is essentially a passive process, the respiratory process is initiated by inspiratory inhibition (Hering-Breuer reflex). An elevated CO_2 pressure prolongs this reflex time, resulting in an increased depth of inspiration.

As a result of endurance training, respiratory frequency at rest and during exercise is reduced, whereas tidal volume is increased (p 294).

4 Effect of Environmental Factors on the Physiology of Work Performance

4.1 Effect of Acute and Chronic Oxygen Deficiency on Man

Oxygen is absolutely necessary for aerobic metabolism because it is the substance used to maintain all physiologic reactions over prolonged periods of time. A complete absence of oxygen is compatible with life for only a short time. A partial withdrawal produces momentary reactions and chronic adaptations which make it possible to survive or even to maintain basic functional capacity. Within the framework of this theme, we will be occupied more with those oxygen deficiencies which affect the organism from without and less with those which appear within the body as a result of circulatory insufficiency or heart failure.

Appropriately, three conditions are differentiated: sudden acute (peracute), acute and chronic oxygen deficiency. Peracute oxygen deficiency appears with a sudden drastic withdrawal of oxygen, such as is found in drowning accidents. Acute oxygen deficiency occurs at high altitude; this plays an important role in mountain climbing and sport flying. Chronic oxygen deficiencies primarily produce problems of work capacity and adaptation in man when he has to work or when he takes part in competition at high altitude. Since the experience of the Olympic Games in Mexico City in 1968, altitude training has been applied more or less successfully.

4.1.1 Sudden Acute Oxygen Deficiency

The amount of oxygen in the body is illustrated in Figure 102. Arterial blood found in the pulmonary veins, left side of the heart and arterial system contains about 280 ml, whereas on the average another 600 ml is contained in venous blood. The muscle pigment myoglobin binds about 240 ml and the lungs at the mid-breathing position contain another 370 ml. At a resting oxygen requirement of 300 ml/minute, it is theoretically possible to use the total amount of oxygen in 5 minutes. Of course, O_2 pressure cannot drop to zero.

Experimental investigations of oxygen pressure after differing periods of apnea at rest are shown in Figure 103. The P_{O_2} in the alveoli drops somewhat, reaching a level of 30 mm Hg in the 3rd minute. This drop is even faster during exercise. If the P_{O_2} falls to less than the limit of

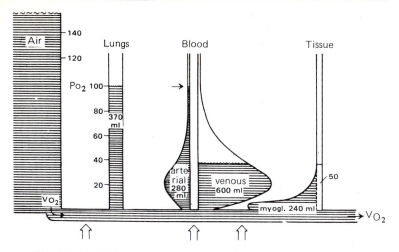

Fig 102   Model of oxygen reserves in man. Oxygen is represented as a fluid, whose height corresponds to the partial pressure in each reservoir. Partial pressure of gas is controlled by valves denoted by the arrows. From Rahn, In: Muskelstoffwechsel, ed. by Keul, Doll and Keppler (Munich: Barth, 1969).

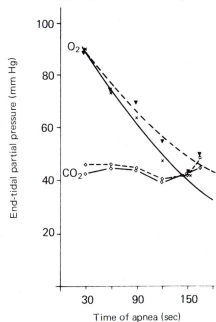

Fig 103
Changes in blood gas
pressures after breath-
holding (solid line, diving;
dashed line, simple
breath-holding).

about 30 mm Hg, then loss of consciousness appears, i. e., there is a marked deterioration in cerebral function. Simultaneously with the drop in O_2 pressure, there is a significant rise in sympathetic tonus; this is especially seen in higher heart rates (tachycardia) and in dilatation of the pupils of the eye (the pupils are dilated via the sympathetic system and constricted via the parasympathetic system). If the oxygen pressure falls even more, then the circulatory center is also implicated. The tachycardia is now transformed into a very slow heart rate (bradycardia). The pupillary center also adjusts its activity, so that the pupils become dilated and no longer react to light. If this continues, there will be stoppage of the heart and respiratory paralysis, followed by clinical death.

At which point in time and under which conditions is survival still possible during this period? It is clear that different organs react differently to oxygen deprivation. During the discussion of muscle physiology, we saw that muscle can work for a considerable time with a local lack of oxygen before it becomes functionally incapable. This is because muscle has a relatively large anaerobic capacity and can cover its energy needs with the energy-rich phosphates and from glycolysis. Therefore, it can go into an oxygen debt. The myocardium and brain tissue, on the other hand, have a poorly developed glycolytic system. As a result, an oxygen deficiency cannot be covered by these reactions for a long time. The cerebral cortex is especially sensitive to a decreased oxygen pressure.

Each organ has a definite survival time, based on its concentration of energy-rich phosphates and its glycolytic activity. Survival time is defined as that time which transpires from the onset of oxygen deprivation to the functional loss of the organ. If oxygen is then resupplied, there will be a return in function after a certain period of recovery; generally, this is longer than the survival time.

Whether lasting damage occurs depends on the time of resuscitation; this is not the same as survival time. At the end of the survival time, there is a cessation of function. During the time of resuscitation, no further irreparable damages to the structure of the cell occur. Thus, with the renewed supply of oxygen, there can be complete restoration in function, even though it requires more time. Only after the period of resuscitation does irreparable damage occur, i. e., cellular death and disintegration (necrosis).

In man, the survival time of the brain is only 8–12 seconds after the sudden local withdrawal of oxygen (e. g., occlusion of blood flow to the brain). The resuscitation time is 8–10 minutes. According to the above definition, there is a loss of consciousness after 8–12 seconds, but up to 10 minutes after a myocardial standstill, the function of the brain may be completely restored. If the resuscitation period is briefly exceeded, then there will possibly be healing with some defects (e. g., intellectual deficiency). The autonomic centers of the medulla oblongata have a longer resuscitation period than the cerebrum.

Although the myocardium is less capable of working anaerobically than skeletal muscle, it does have a somewhat longer resuscitation time than the brain. There is no adequate method to measure the resuscitation time of the total body, but it is basically shorter than that of the brain or heart for the following reasons. As a result of oxygen deprivation, the heart loses so much of its ability to perform that it cannot provide adequate pressure to supply blood to the brain. The resuscitation of the brain occurs only when sufficient amounts of oxygen are being delivered. This is the main reason the work of the heart must be externally supported during the period of resuscitation. All measures aim first at inducing the autonomic centers to begin working spontaneously; this circulatory support is produced by the mechanical compression of the heart (heart massage) and by the periodic renewal of the contents of the lungs (artificial respiration). During this time the airways must be free of obstruction. If resuscitation is successful, then the autonomic controls are the first to be reestablished. The work of the heart must usually be supported until the patient regains consciousness.

4.2 Relationships between Altitude and Oxygen Pressure

The oxygen concentration of the earth's atmosphere (dry) is uniform and has a value of 20.95%. The partial pressure of oxygen at sea level is therefore $(20.95 \cdot 760)/100 = 159$ mmHg.

This pressure decreases with increasing altitude, corresponding to a drop in barometric pressure, so that at an altitude of 5,500 m it has dropped by one-half. The precise data on the relationship of altitude and atmospheric oxygen pressure is shown in Figure 104.

Several remarks about this diagram are necessary. The ordinate on the right lists barometric pressure: sea level corresponds to a pressure of 760 mm Hg equal to 760 Torr equal to 1,000 mb equal to 1 atmosphere. On the left ordinate is the partial pressure of oxygen at a relative humidity of 0. In actuality, air found in the alveoli and airways is 100% saturated with water vapor. Water vapor pressure at complete saturation is governed by temperature, which in this case is identical to a body temperature of 37 C. At 37 C, water vapor pressure is equal to 47 mm Hg; this must be subtracted from barometric pressure before partial pressure is calculated. Thus, at 760 mm Hg and complete saturation, the partial pressure of oxygen is 20.95% of (760–47) or 149

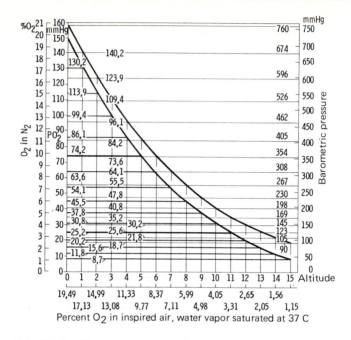

Fig 104 Relationship among barometric pressure, altitude and partial pressure of oxygen in dry (upper curve) and humid (lower curve) air. From Pichotka, J.: Der Gesamt-Organismus im Sauerstoffmangel, In: Handbuch der allgemeinen Pathologie, vol. IV-2, 395 (Berlin: Springer, 1957).

mm Hg. From the diagram, it is possible to read which O_2 pressure corresponds to which altitude (upper abscissa in km); the upper curve is for dry air and the lower curve is for water-saturated air at 37 C. The numbers located on the lower curve simplify reading O_2 pressure values at different kilometers of altitude. The lower abscissa represents the per cent mixtures of O_2 and N_2 needed under normoxic conditions to simulate the corresponding altitude. As an example, the O_2 pressure under saturated conditions at an altitude of 4,000 m is 86.1 mm Hg. The same O_2 pressure can be produced at sea level with a mixture of 11.33% O_2 in N_2 because 11.33% of 760 mm Hg is also 86.1 mm Hg.

To correctly classify the effects of altitude, refer to Figure 105, in which the altitude is similarly compared to actual barometric pressure. The fraction of respiratory gases in the lungs is schematically depicted as partial pressure. The water vapor pressure at sea level is 47 mm Hg, CO_2 pressure is about 40 mm Hg, O_2 pressure is 100 mm Hg and the rest is related to the partial pressure of nitrogen. Although ventilation

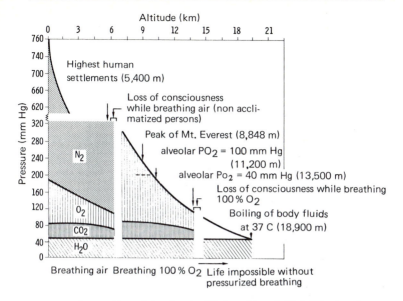

Fig 105   Composition of alveolar air while breathing air (0–6.000 m) and pure O_2 (6,000–13,500 m) at environmental pressure. From Ganong, W. F.: Medizinische Physiologie (Berlin: Springer, 1971).

increases due to the drop in O_2 pressure (p 162), P_{O_2} falls to a critical value of 30 mm Hg at an altitude of about 6,000 m; this causes a functional failure of the cortex. As a result of the hyperventilation caused by hypoxia, there is a reduction in CO_2 pressure. Altitudes greater than 6,000 m cannot be tolerated by persons unacclimatized to altitude (p 174). With breathing masks supplying pure oxygen, altitudes up to 13 km, can be tolerated. After that point, pressurized cabins are needed for survival. Modern commercial airplanes fly up to altitudes of 11 km but they have compressors which bring the cabin pressure to that corresponding to about 2,000 m. If these cabins were not airtight, then survival would not be possible. For this reason, they must be equipped with oxygen masks which will supply pure O_2 to the passengers in case of decompression.

4.2.1 Acute Oxygen Deficiency

The first symptoms of acute hypoxia, also known as altitude or mountain sickness, are shortness of breath and heart palpitations, producing apathy, paleness and the onset of sweating. Characteristic

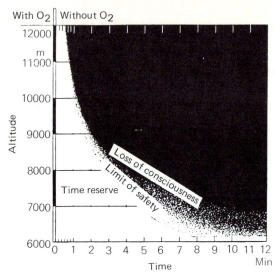

Fig 106   Effect of altitude after interruption of breathing oxygen at altitudes of 7,000–12,000 m. From Ruff-Strughold: Grundriss der Luftfahrtmedizin (Munich: Barth, 1957).

psychic changes also occur. The earliest symptoms are euphoria, an increased sense of well-being and an overestimation of one's ability. After that, the person becomes injudicious, quarrelsome and then depressed and unresponsive. This picture resembles that found during the different stages of drunkenness. Mountain sickness first appears at 3,000 m, although disturbances can be seen at 1,800 m in labile and older persons.

After a sudden drop in oxygen supply, there is a certain time reserve (Fig 106), the magnitude of which depends on the altitude. Due to the awareness of their time reserve, parachutists and sports pilots still have the possibility of reaching high altitudes without having symptoms of altitude sickness or loss of consciousness.

The effects of oxygen deficiency have also been studied in low-pressure chambers. Figure 107 shows a writing test after ascending to an altitude of 7,500 m. The psychophysical performance obviously becomes worse with time, i. e., the subject is no longer able to write a clear sentence. The primary danger of acute oxygen deficiency and the resulting mountain sickness is that an individual's own work capacity will be falsely estimated and incorrect reactions may cause a fatal outcome.

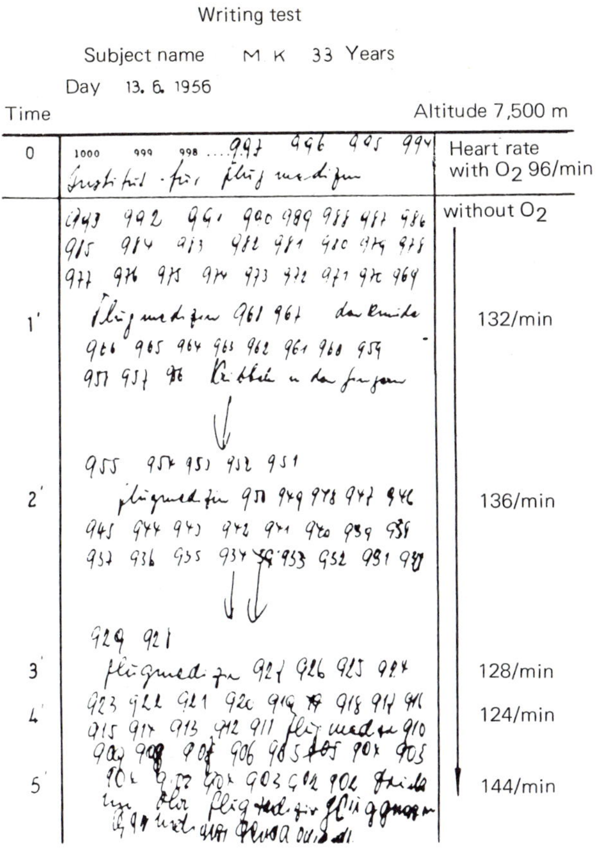

Writing test

Subject name   M K   33 Years

Day   13. 6. 1956

Time   Altitude 7,500 m

Time		Heart rate
0	1000 999 998 ....*997* *996 995 994*	Heart rate with O$_2$ 96/min
	[handwritten text]	
	[handwritten numbers]	without O$_2$
1'	*[handwritten text]*	132/min
2'	*[handwritten text]*	136/min
3'	*[handwritten text]*	128/min
4'	*[handwritten text]*	124/min
5'	*[handwritten text]*	144/min

Fig 107   Writing test at altitude of 7,500 m after suspension of breathing oxygen. From Ruff-Strughold: Grundriss der Luftfahrtmedizin (Munich: Barth, 1957).

4.2.2 Chronic Oxygen Deficiency and Altitude Acclimatization

Peracute and acute oxygen deficiency will always remain exceptional situations, whereas all who work, live and play for various periods of time at altitude are exposed to chronic oxygen deprivation. No difficulties are generally expected in healthy persons up to an altitude of 2,000 m. Tolerance and physical working capacity at high altitude depend primarily on the duration of residence there, because the organism has a number of adaptive mechanisms at its disposal; these are summarized by the collective term "altitude acclimatization."

The first phase of altitude adaptation occurs immediately after arrival and consists of increased ventilation and cardiac output. Ventilation is controlled by the effect of the drop in O_2 pressure on the arterial chemoreceptors located in the carotid and aortic bodies. Animal research, in which the supplying nerve was cut, indicated that the effect of oxygen deprivation on ventilation is increased, whereas the effect on the circulation remains unchanged. The conclusion is that heart rate and ventilation are controlled by different systems during periods of oxygen deficiency. It is likely that the muscle receptors are switched on to help regulate heart rate (p 133). The effect of the first phase of altitude adaptation is such that the reduced oxygen pressure caused by a drop in barometric pressure is partially compensated by increased ventilation. The rise in cardiac output partially balances the reduced saturation so that the cells will still be supplied with O_2.

Of course, this compensation is insufficient to counteract the effect of a lack of O_2. Man's physical work capacity must therefore be considerably reduced, especially because the energy requirements for the respiratory and circulatory systems are greater than those found at sea level, i. e., a portion of their reserves is already required. In this regard, there is a further detrimental effect. The increased ventilation simultaneously produces a situation where P_{CO_2} and $[H^+]$ become less, resulting in a shift to the left of the oxygen dissociation curve. Thus, we have a classic example of respiratory alkalosis. This means that the oxygen pressure at the cellular level is less for the same saturation. A part of the effect of an increase in ventilation is therefore lost. The reduced P_{CO_2} and $[H^+]$ also cause a part of the additional stimulus resulting from the drop in P_{O_2} to be compensated by the reduction in the other two respiratory stimuli. The effect of a drop in P_{O_2} on ventilation is shown in Figure 98.

The second phase of altitude acclimatization has a double effect; this phase also begins on the initial day at altitude. Firstly, there is an increase in the content of 2,3-DPG in the erythrocytes. As we have already seen on page 94, the effect of this substance is to shift the O_2 dissociation curve to the right. In this way, the effects of reduced $[H^+]$ and P_{CO_2} on the curve are compensated. Secondly, there is a reduction in the concentration of bicarbonate because the alkali sparing mechanism of the kidney is inhibited, i. e., increased levels of alkaline ions are now excreted by the kidneys. The effect of this compensation is illustrated in Figure 90 D. The CO_2 dissociation curve becomes flatter and as a result the same CO_2 pressure produces a higher hydrogen ion concentration. Ventilation then increases, again causing an elevation in oxygen pressure in the alveoli. The energy needs of breathing are

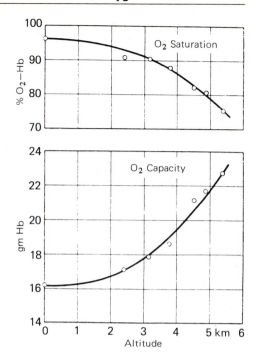

Fig 108
Relationship between
O_2 saturation and O_2
capacity with altitude
acclimatization at
4300 m. From
Pichotka, J.: Der
Gesamtorganismus
im O_2-Mangel. In:
Handbuch der all-
gemeinen Pathologie,
vol. IV-2, 3590 (Ber-
lin: Springer, 1957).

now increased as a result of the higher ventilation, but the performance reserves of the body are not appreciably improved. On the other
hand, oxygen pressure at the cellular level is higher once more.

The third phase of altitude acclimatization consists essentially of the
gradual increase in blood volume, erythrocyte count and hemoglobin
content. This is a sluggish process and it is still not completely clear
how much time is required for this adaptation. Evidently, there are a
number of mechanisms participating in this process, the details of
which have not been clarified. Acclimatization should occur faster if
the person had ever acclimated to altitude before (tissue factor according to Opitz) or if the person was born at altitude. Our knowledge of
the time course comes basically from the results of expeditions, where
several other factors such as training, physical work capacity or radiation are also present during those extreme conditions.

There are especially good conditions for studying the problems discussed above in the Peruvian Andes because there are settlements up
to an altitude of 5,300 m. It has been instructive to investigate the
permanent residents living at different altitudes and to compare their
data with those of men living at sea level in the capital city, Lima.

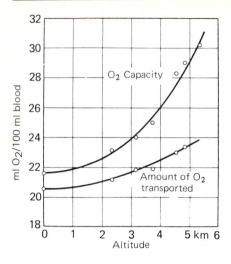

Fig 109
Oxygen capacity and amount of O_2 transported at different altitudes in acclimatized men. From Pichotka, J.: Der Gesamtorganismus im Sauerstoffmangel. In: Handbuch der allgemeinen Pathologie, vol. IV-2, 3950 (Berlin: Springer, 1957).

Figure 108 shows some of the findings. As expected, the O_2 saturation of arterial blood is less with increasing altitude. Beginning at an altitude of 3,000 m, the saturation falls off more quickly. At the highest settlement (Quilcua at 5,340 m), an average arterial saturation of only 76% was observed. The higher the altitude where man can live, the greater the oxygen capacity of the blood. Both figures show that the O_2 capacity is essentially influenced by the arterial saturation. At an altitude of 4,900 m, the O_2 capacity is about 34% higher than that at sea level; it is estimated to be 45% higher at 5,300 m. As a result of this increase, there is a compensation for the reduced saturation. The O_2 transport capacity is overcompensated; there is a rise in the amount of O_2 transported with increasing altitude because the effect of an increased hemoglobin content is greater than that of a reduction in per cent saturation. For this reason, O_2 content of blood is therefore probably not the controlled variable.

This fact is also seen in Figure 109, where the amount of O_2 transported is recorded. This is calculated as the transport capacity (corresponding to 100% saturation) and the actual saturation. In the entire range studied, the transported amount of O_2 rises disproportionately with altitude. Figure 110 illustrates the effects of adaptation. Newcomers to altitude who have not yet experienced the second and third phases of altitude acclimatization, have a considerably lower O_2 saturation than permanent residents.

The limits of adaptation are at about 5,000 m, where limiting values of 8,000,000 erythrocytes and 25 gm hemoglobin per 100 ml occur after complete acclimatization.

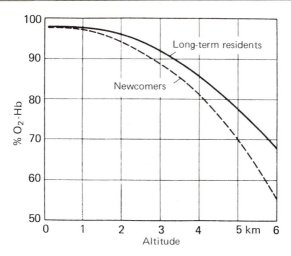

Fig 110   Oxygen saturation as function of altitude in longterm residents and in newcomers. From Pichotka, J.: Der Gesamtorganismus im Sauerstoffmangel. In: Handbuch der allgemeinen Pathologie, vol. IV-2, 3950 (Berlin: Springer, 1957).

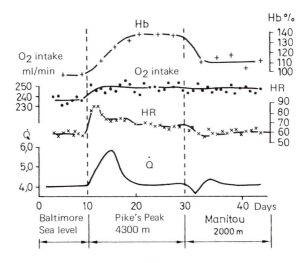

Fig 111   Alterations in hemoglobin (Hb), oxygen intake, heart rate (HR) and cardiac output (Q̇) over acclimatization period of 20 days at 4,300 m. From Pichotka, J.: Der Gesamtorganismus im Sauerstoffmangel. In: Handbuch der allgemeinen Pathologie, vol. IV-2, 3950 (Berlin: Springer, 1957).

Relative to the control aspects, there are the following interesting observations obtained during prolonged residency at altitude (Fig 111). In the first and second phases of adaptation, there is a marked rise in cardiac output and heart rate. As soon as the transport capacity is increased, both variables then fall to or even under their initial values. Ventilation (not shown) and its associated O_2 requirement remains elevated during the entire time. Thus, it is confirmed that ventilation is evidently controlled by the effect of reduced P_{O_2} on chemoreceptors. There is little improvement in P_{O_2} during adaptation to altitude. The circulatory variables are more related to the energy content of the muscles with their poorer supply caused by oxygen deficiency, when the transport capacity for oxygen has adapted to altitude.

4.2.3 Work Capacity during Oxygen Deficiency

During the 1968 Olympic Games in Mexico City, there was the opportunity to conduct a number of investigations on physical work capacity at an altitude of about 2,500 m. Due to the nature of the Games themselves, the subjects were primarily well-trained athletes. As a general trend, most performances of brief duration were unchanged or even improved, whereas those of long duration were basically worse. This finding is logical because the energy needs of brief, high-level performance are essentially anaerobic and those of prolonged performance are aerobic (p 49). The improved performance, especially in the speed disciplines (e. g., cycling, running), is related to the reduction in wind resistance. The pressure that a gas exerts on a surface is directly proportional to the number of air molecules and their kinetic energy. Because the pressure of impact rises in proportion to the square of the speed, it is less at altitude. In comparison to sea level, the greater the speed, the more noticeable is this reduction. This hypothesis is also confirmed by the fact that swimming performances over short distances were essentially identical at altitude and at sea level.

The quantitative changes in running performance can be seen in Figure 112. All world records existing before the 1968 Olympic Games in Mexico City are indicated by 100% on the vertical midline, with the various distances indicated. The corresponding record times are given in parentheses alongside the actual times recorded in Mexico City. There is a clear improvement in performance at the shorter distances and a deterioration at the longer distances.

Physiologically, lactate production during endurance events was greater than at sea level. Likewise, there was a greater reduction in blood glucose levels, reflecting the greater fraction of anaerobic energy production.

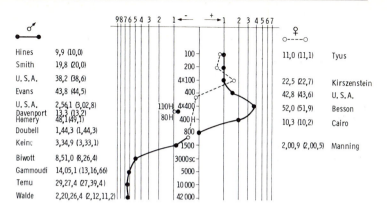

Fig 112   Deviations in records at Olympic Games in Mexico City compared to world records achieved at sea level. Ability to perform at altitude decreases in endurance events and increases in brief events. From Jokl, E.: Bericht über die sportärztlichen Untersuchungen bei den Olympischen Spielen in Mexico City 1968. 5. Gymnaestrada Basel, Wiss. Symposium: Basel, 1969.

It is interesting that work allows man to reside at high altitudes at which he can no longer live at rest. The highest settlement in the Andes is at 5,300 m. The mines associated with this settlement are several hundred meters higher. Attempts to get the men to settle directly at the site of the mines failed due to sleeplessness, loss of appetite and a reduction in nutritional intake. It appears that the working stimulus to the muscle receptors is necessary to maintain high levels of ventilation, so that the P_{O_2} in alveolar air does not drop below a critical level. The stimulation due to an oxygen deficiency alone is insufficient at rest.

Injuries, especially to the heart, were not observed under hypoxic conditions, even during high-performance activities. Apparently, skeletal musculature and the nervous system give up sooner, so that work must be interrupted due to exhaustion before injuries develop. On the other hand, such functional disturbances as collapse and migraine headaches are frequent.

4.3 Influence of Climate on Man

Climate is that condition characterized by environmental temperature, humidity of the air, heat radiation and wind velocity. Dependent on the geographic location and on the season of the year, climate is so variable that there are zones where man cannot live, despite civilizing

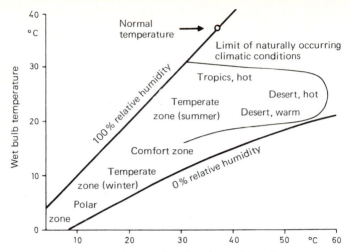

Fig. 113   Range of naturally occurring climatic conditions (temperature and humidity). From Wenzel, H. G.: Die Wirkung des Klimas auf den arbeitenden Menschen. In: Handbuch der gesamten Arbeitsmedizin, vol. 1 (Berlin: Urban and Schwarzenberg, 1961).

devices, as well as zones where these are practically not needed because man's biologic regulatory processes are sufficient. Between these two extremes, man must live and work (Fig 113). Our task will be to consider man's capacity to work and live in this range, which can still be greatly varied by technical means.

It has been known for some time that temperature alone is not the decisive factor which affects one's ability to work and his comfort, and that other factors must be responsible. Such factors as relative humidity, wind velocity and heat radiation, i. e., factors that can be determined accurately enough with the proper apparatus, must also be considered. It is more difficult to judge the value of these individual factors relative to well-being and performance capacity.

As early as the turn of the century, it was recognized that air temperature measured with a dry thermometer (dry bulb temperature) had to be evaluated differently according to the vapor content of the air. Temperature recorded on a dampened thermometer gives a better comparison with temperature sensation. The temperature obtained from a thermometer encased in a damp cloth is called the wet bulb temperature. Depending on the relative humidity, the thermometer is affected by the cold produced by evaporation; the wet bulb temperature is the same as the dry bulb temperature only at 100% relative humidity. Wet bulb temperature is only an incomplete measure of

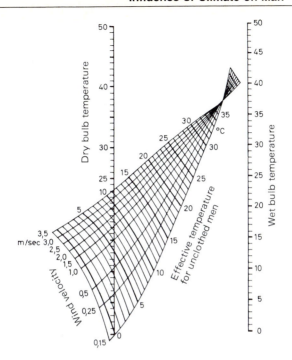

Fig 114   Effective temperature for unclothed men in relation to dry bulb temperature, wet bulb temperature and wind velocity (according to Yaglou).

climate and its effect on man. For this reason, man has looked for a definite index of climatic condition based on temperature, humidity and wind velocity. There were then studies to determine which combination of the three factors produced the same temperature sensation in resting man, e. g., the sensation of the same temperature was produced at zero wind velocity relative to a series of variations in relative humidity. This relative variable is called the effective temperature. It is therefore not a basic physical variable, but a variable based on sensations. It is obvious that this compound climatic index must produce different values for a clothed person than for an unclothed person.

Figures 114 and 115 present the diagrams needed for the determination of effective temperature for unclothed and clothed man.

The diagrams are read in the following manner. The value of the dry bulb temperature on the left vertical line is connected by a straight line to the value of the wet bulb temperature on the right vertical line. The intersection of this line with the value of the wind velocity then

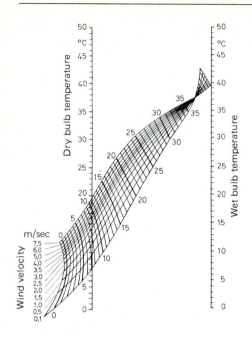

Fig 115
Effective temperature for clothed men in relation to dry bulb temperature, wet bulb temperature and wind velocity (according to Yaglou).

yields the value of the effective temperature. If an unclothed person encounters a dry bulb temperature of 30 C, a wet bulb temperature of 20 C and a wind velocity of 1 m/second, the effective temperature is 22 C. The same temperature sensation would be felt by the unclothed person at 22 C dry bulb temperature, 100% humidity and zero wind velocity.

However, this effective temperature does not take into consideration the effect of radiation. Man has endeavored to find other combined measures of climate in which this factor is also considered.

4.3.1 Heat Exchange between Body Surface and the Environment

4.3.1.1 Heat Exchange via Conduction and Convection

The surface of the body can exchange heat with the environment in different ways. Heat exchange with the environment occurs primarily due to heat conduction. This form of heat exchange is described by the following equation:

$$\dot{Q}_K = S_e \, \lambda \, (t_s - t_m) \text{ kcal/minute,} \qquad (58)$$

where λ is heat conductivity and S_e is the effective body surface area involved in the exchange; t_s is surface temperature and t_m is the temperature of the surrounding medium. Pure conduction has only a minor role in heat loss during occupational work and during sport and occurs only when a solid object comes into contact with the skin surface. Nevertheless, a considerable amount of heat is given off, for example, when one stands barefoot on a cold surface.

In the presence of fluid or gaseous (e. g., air) media, an additional effect occurs which supports heat conduction, i. e., convection (C). Due to the relationship between the weight of air and temperature and due to the diffusion of air molecules, air is always moving over the skin surface, as long as a temperature gradient exists between the skin and air. For practical purposes, we differentiate this "free convection" from "induced convection," which occurs when air is forced to move against the skin, whether by naturally occurring air currents or by a fan. Practically speaking, because heat conduction and convection belong together, the formula for heat loss by conduction and convection is written as follows:

$$\dot{Q}_{KC} = S_e \, h \, (t_s - t_m) \text{ kcal/minute.} \qquad (59)$$

h in this form refers to the heat transfer coefficient, which in turn is composed of two parts, the effect of the heat conduction coefficient and the effect of convection. It is generally of the magnitude of 5–9 $\text{kcal} \cdot \text{hr}^{-1} \cdot \text{m}^{-2} \cdot \text{C}^{-1}$. The heat conduction coefficient of tissue amounts to $0.75–3.5 \cdot 10^{-3} \text{ cal} \cdot \text{cm}^{-1} \cdot \text{sec}^{-1} \cdot \text{C}^{-1}$ and is thus about 10 times greater than that of cork and 1,000 times smaller than that of silver. It indicates what amount of heat (cal) can penetrate through a 1 cm thick layer at a temperature difference of 1 C/second. The temperature gradient is maintained by convection. Without it, the layer of air lying over the skin would be warmed to the temperature of the skin and there would no longer be a temperature difference. Free convection is less in water than in air.

Thus, if one remains motionless in water at a temperature of 18 C, then he heats the water in the immediate vicinity of his skin and reduces the temperature gradient between the blood and the surface. If he swims, then induced convection maintains the temperature gradient between his blood and his skin and the heat loss will generally increase more than the corresponding heat production due to swimming. As a result, the body cools off (for more details, see p 227).

4.3.1.2 Heat Exchange via Radiation

All objects mutually irradiate each other with energy. In this way, warmer objects are cooled and colder objects are warmed. This physical law is also followed by the human body, whose emitted radiation energy obeys the Stefan-Boltzmann Law, which says that the total amount of heat energy radiated from the surface S per unit of time ($\dot{Q}_R$) is proportional to the temperature to the power of 4 (measured in degrees Kelvin):

$$\dot{Q}_R = \varepsilon \, \sigma \, S_e T_s^4 \qquad (60)$$

(ε = emission coefficient, σ is the radiation constant $4.96 \cdot 10^{-8}$ kcal $\cdot$ hr$^{-1} \cdot$ m$^{-2} \cdot$ degree Kelvin^{-4}).

This law describes the heat energy which is taken from or added to an object via radiation. For this reason, the simultaneously absorbed amount of heat per unit of time must be subtracted; this is primarily governed by the surface temperature of the radiating bodies in the environment (T_m). The absorbed amount of heat is then:

$$\dot{Q}_a = \delta \, \sigma \, S_e \, T_m^4. \qquad (61)$$

(δ = absorption coefficient)

The difference in heat flux is thus:

$$\dot{Q}_f = \sigma \, S_e \, (\varepsilon T_s^4 - \delta T_m^4). \qquad (62)$$

The wave lengths λ radiating from the body are well within the range of infrared ($\lambda = 3$–60 µm). Therefore, the skin radiates in a manner similar to that of a black body which has an emission coefficient of 1. This emission coefficient is not dependent on the color of the skin. The absorption coefficient of the most important naturally occurring heat radiation (solar radiation, whose wave lengths are between 0.3–4 µm) is dependent on the pigmentation of the skin. The absorption coefficient of whites lies between 0.6 and 0.7 and extends to 0.8 in blacks.

4.3.1.3 Heat Exchange via Evaporation and Condensation

All evaporating fluids take energy from their environment in the form of heat. The amount of heat in kcal withdrawn per liter fluid is called the specific heat of evaporation. All condensing fluids give off the same amount of heat per liter fluid as that required for evaporation (E).

The human body evaporates primarily water, which has a specific heat of evaporation of 580 kcal/l.

The heat exchange via evaporation and condensation can be described by the following equation:

$$\dot{Q}_E = S_e \beta \, (P_s - P_m), \qquad (63)$$

where β is the evaporation coefficient, P_s is the vapor pressure of the surface and P_m is the vapor pressure of the surrounding air. The evaporation coefficient is naturally dependent on convection, just as has been described for the heat transfer coefficient, because it can modify the humidity gradient between the skin and the surrounding air.

4.4 Body Temperature and Heat Balance

There is neither a locally nor a temporally uniform body temperature among the so-called homeothermic organisms to which man belongs. There is no temporal uniformity because body temperature is also subject to circadian rhythms (p 234). There is no local uniformity because there is a continual flow of heat from the center toward the periphery. This flow of heat can occur only when there is a temperature gradient within the body.

Therefore, the essentially constant core temperature and the more variable shell temperature are distinguished; the limits of each are not anatomically fixed but are shifted functionally, depending on the environmental conditions to which the body is subjected. Figure 116 shows schematically the patterns of the isotherms within the organism with exposure to low or high external temperatures. Seventy per cent

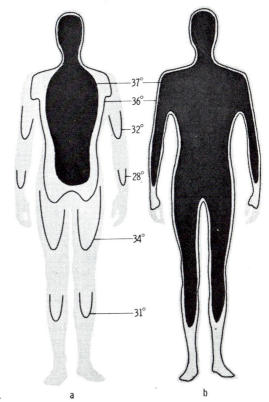

Fig 116
Isotherms in man at low (A) and high (B) external temperatures (according to Aschoff and Wever).

a   b

of the total amount of heat produced by the basal metabolism is generated in the brain, as well as in the thoracic and abdominal organs. This region is called the body core. The body shell is composed primarily of the extremities, which have an especially important function in temperature regulation.

Both the more commonly measured rectal temperature and the oral temperature show values which incompletely reflect core temperature. However, for practical purposes, they are generally sufficient.

The heat content of the body (Q_b) is determined by the amount of heat produced plus the amount taken in from the outside and by the amount of heat given off. Heat balance can be determined from uptake, production and output of heat. Under stationary conditions, heat content is constant when the balance is in equilibrium. If we state this balance mathematically, then the following formula will be produced:

$$\Delta Q_b = (\dot{Q}_M \pm \dot{Q}_{KC} \pm \dot{Q}_f \pm \dot{Q}_E) \cdot t, \tag{64}$$

where ΔQ_b represents the change in body heat content, $\dot{Q}_M$ the metabolic heat produced per unit of time and t the time; the other indices of $\dot{Q}$ (K, C, f and E) have been discussed in the previous section.

4.4.1 Heat Production

Due to basal metabolism alone, adults generate about 1.1–1.4 kcal/minute. When man works, the amount of heat generated increases rapidly. If the working metabolism and the rate of efficiency (given as η in %) of the work load are known, the additional amount of heat generated per unit of time ($\dot{Q}$) is:

$$\dot{Q} = \frac{\text{Working metabolism } (100-\eta)}{100} \text{ kcal/minute.} \tag{65}$$

An athlete with a maximal oxygen intake of 5 l/minute while exercising on a bicycle ergometer ($\eta = 20\%$) thus produces 20 kcal/minute. If we can visualize a situation where the body suddenly can no longer give off heat, then the temperature would rise, depending on the specific heat, body mass and time. The average specific heat of the human body is 0.8 kcal/kg · degree C. Thus, 0.8 kcal is needed to increase the temperature of 1 kg tissue by 1 C. A 70-kg man under basal conditions would increase his body temperature 1 C in about 1 hour. With a working metabolism of 20 kcal/minute, his body temperature would rise 1 C in only 3 minutes. If body temperature is to remain constant during exercise, then just as much heat must be given off as is produced.

Of course, such calculations oversimplify the actual relationships. Because the body shell is temperature variable to some extent, the total heat content of the body can be modified to a considerable degree without causing a rise in core temperature. If we assume that the body shell (roughly one-third of the body mass) with a mean temperature of 34 C increases its temperature to 40 C, as is actually the case during exercise, then there can be a rise in total heat content of about 130 kcal, without producing an elevation in core temperature.

While heat production in an exercising man is a by-product of the mechanical energy produced, a resting man in a cold environment will use it to regulate his core temperature. This heat is also formed in the skeletal muscles, either by an elevation of muscle tone or if a high level of heat needs to be produced, by an asynchronous stimulation of the muscles, producing shivering.

4.4.2 Heat Loss

The physical fundamentals of heat exchange with the environment have already been discussed. The magnitude of the individual heat fluxes depends primarily on climatic and environmental conditions and can be estimated or calculated only for special conditions. Figure 117 schematically shows the fractional heat loss due to radiation, conduction, convection and evaporation when heat production is constant. Particularly consider the effect of heat radiation, because empirically it is easy to overlook.

The figure should make the proportions present in an interior room at the place of work more graphic when the external temperature is extremely low. If the screen placed before the window with a temperature of 2 C did not obstruct heat radiation, the worker at rest would be chilled, despite the air temperature of 24 C, because he would radiate

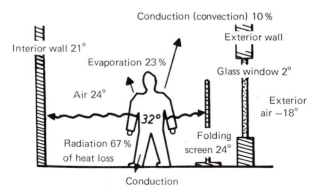

Fig 117   Proportion of heat loss associated with radiation, conduction and evaporation in resting man under defined climatic conditions (according to Gillies).

too much heat. One can again and again observe patients in a hospital being prepared for an operation who are left near cold windows. This is due to the mistaken belief that a room temperature of 25–26 C is sufficient to keep the patients from cooling off.

Light, airy gymnasia with large windows are preferred in modern sports facilities. The physical education teacher must remember that air temperature in the gymnasium or swimming pool is not the only decisive factor relative to comfort and that a large amount of heat will be radiated to the window surfaces when it is cold outside. Cooling and those diseases associated with it can then result.

Adjustment of heat loss in man is produced by altering blood flow to the skin and by secreting sweat. Because the heat conductivity of skin is low, the effect of a modification in blood flow is that it alters the amount of heat lost to the skin. In addition, skin temperature varies with its blood flow. The greater the flow of blood through the skin, the more the temperature of the skin will approach that of the blood. The skin temperature on the body's shell, which can be extremely low in a resting man exposed to cold temperatures, can increase more than 15 C. Thus, the loss of heat via conduction and radiation is greatly increased.

Because the effective body surface involved is a factor in all of the heat exchange formulas, it is easy to see how the extremities have an especially important role in heat loss. Due to the geometric structure of the extremities, their ratios of surface:volume are larger than that of the trunk. Also, blood flow under physiologic conditions can be altered to a much greater extent to the extremities than it can to the trunk. For example, blood flow to the hand can vary at a ratio of 30 : 1, whereas that to the fingers can vary at a ratio of 600 : 1. Considering the pressure gradients present, such extreme alterations in blood flow can occur only when the arteriovenous anastomoses are opened or closed. The initiating mechanism can be either a direct temperature effect or a central nervous effect.

Heat loss due to evaporation at normal room temperatures and complete physical rest occurs in the respiratory tract and also because skin is not completely watertight. Depending on the temperature, 20–40 gm water per hr penetrates and then evaporates as insensible perspiration. Under these conditions, sweat can still be secreted by the glands located on the forehead, inner surfaces of the hands and under the armpits, primarily due to psychic emotions. Skin temperatures above 35 C initiate the secretion of sweat (sensible perspiration). The fibers initiating the secretion originate in ganglia of the sympathetic nerve and go to the sweat glands. The production of sweat is an active performance of these gland cells. Even skin with low levels of blood flow can perspire. The fluid in sweat naturally comes from the blood.

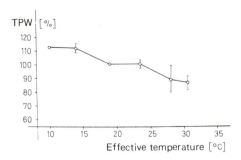

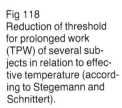

Fig 118
Reduction of threshold
for prolonged work
(TPW) of several sub-
jects in relation to effective temperature (accord-
ing to Stegemann and
Schnittert).

Heat loss during physical work needs a few more supplementary
explanations. Heat produced in muscle is essentially eliminated
through the skin. The circulatory system, which represents the weakest
link in the chain during exercise, is therefore stressed more when the
possibilities of heat loss are more unfavorable. During work with small
muscle groups, where muscle blood flow determines the threshold for
prolonged work, muscle and skin often are supplied by the same
arteries. This means that under unfavorable heat conditions, the
supplying artery becomes the limiting factor for maximal blood flow
and the threshold for prolonged work decreases. In the case of exercise
with large muscle groups, the performance capacity of the heart may
be overloaded by the large blood flow to the skin; this will also produce
a reduction in the threshold for prolonged work. If a large amount of
sweat is lost, there is additionally an increase in blood viscosity and a
transient reduction in blood volume, resulting once more in an
impaired supply to the muscles. Figure 118 shows the association
between the threshold for prolonged work and climate. It can be
clearly seen that the threshold drops with a rise in effective temperature.

The manner in which the body eliminates its produced heat depends
primarily on external factors. Figure 119 shows a typical example at a
temperature of 37 C and during physical work at different levels of
energy metabolism. There is a disproportionate rise in sweat production in relation to the increase in energy metabolism. Measurements of
sweat loss are shown in the upper portion of the figure. The diagram
shows that the increase in sweat production is greater than the increase
in heat production. As a result, there is a drop in mean skin temperature, as shown in the lower portion of the figure. This drop in surface
temperature means that there will be a reduction in the temperature
gradient relative to environmental temperature and a lower value for
heat loss via conduction, convection and radiation. The physiologic
importance of this change in heat loss is the accompanying augmentation of the internal temperature gradient between the core and surface

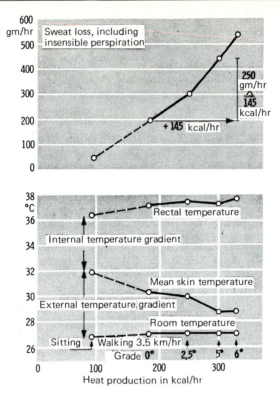

Fig 119   Sweat loss, rectal temperature and mean skin temperature of man at different levels of heat production. Room temperature, 27 °C; relative humidity, 60–65%; air movement, 0.5 m/second. From Wenzel, H. G.: Die Wirkung des Klimas auf den arbeitenden Menschen. In: Handbuch der gesamten Arbeitsmedizin, vol. 1 (Munich: Urban and Schwarzenberg, 1961).

of the body. This augmentation spares the need for an elevated skin blood flow normally caused by increasing work intensity; this is not always easy to provide because the working muscles also need more blood. Sweat production is initiated during exercise not only to the extent required for heat balance, but also so that the thermal tasks of the circulatory system will be facilitated. Figure 120 is a summary of those room temperatures at which an unclothed person performing varying intensities of work will not produce any sweat. The line drawn in the diagram can be considered the comfort temperature for the given work load and a consequence of the varying levels of energy metabolism at the given room temperatures. Those combinations

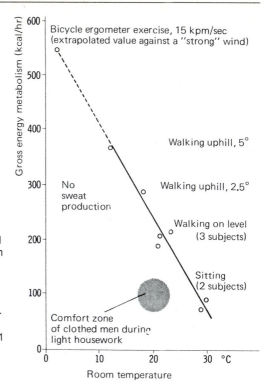

Fig 120
Maximal possible energy metabolism (without sweat production) of unclothed men at different room temperatures. From Wenzel, H. G.: Die Wirkung des Klimas auf den arbeitenden Menschen. In: Handbuch der gesamten Arbeitsmedizin, vol. 1 (Munich: Urban and Schwarzenberg, 1961).

above and to the right of the line, at which sweat is produced, would be considered warm; the combinations conforming to the pattern of the line are considered thermally neutral. From an objective standpoint, environmental temperature can be called comfort temperature when there is neither an increase in metabolism (a sign of the chemical heat regulation against cooling), nor production of sweat (a sign of the physical heat regulation against overheating of the body). Therefore, the comfort temperature during physical exercise is a function of the work performed.

In the case of a clothed person, the curve shifts to the left. For the normally clothed person performing work in the office or at home, the comfort zone denoted by the shaded area is valid.

It is repeatedly astounding to see how well the body's heat balance is in equilibrium, even with extreme variations in energy metabolism.

The heat content of the body also changes little with large variations in metabolism. This exceptional constancy of heat content is due to the activity of a precise regulatory system.

4.4.3 Thermoregulation

There are still many questions about the mechanism of thermoregulation and the functions of the centers in man. Apparently, adjustments in heat production and heat loss are produced in different areas of the hypothalamus. The defense reactions to heat originate more from the superior structures; these structures are collectively called the cooling centers. The reactions against cold are initiated more from the inferior area of the hypothalamus; this region is called the heating center. The thermoregulatory centers themselves are thermosensitive. Warming of the cooling center produces peripheral vasodilatation and sweat secretion, whereas cooling of the heating center causes increased metabolism and vasoconstriction.

The thermoreceptors in the skin act in the regulatory process in a manner similar to that of switching on the disturbance input in a technical control system, i. e., they span the slow adjustment time. If one steps into a warm bath, for example, vasodilatation does not wait until the heated blood warms up the hypothalamic region, but occurs as soon as the thermoreceptors are stimulated. The thermoreceptors adapt so rapidly that only the final control is performed via the hypothalamic centers.

Other higher centers also participate in thermoregulation. They direct behavioral thermoregulation; this is the reason a freezing person will curl into a ball in order to reduce the effective surface exposed.

4.5 Acclimatization

Prolonged exposure to a hot environment produces adaptations in the organism which permit man to better resist this climate. The adaptive processes are especially pronounced if the body is also stressed with exercise. On the other hand, there is no real acclimatization to a cold environment. The most important phenomenon of acclimatization is an increased sweat secretion, which can increase by up to one-third of its initial amount during the first 9 days. Figure 121 shows the formation of sweat as a function of rectal temperature during the course of 9 days at a work load of 87 kcal/m^2 · hr, 37.8 C and 90% relative humidity. The rise in sweat production during the course of acclimatization apparently contradicts the general experience that one sweats less during this time. This appearance is deceiving because the sweat loss is more regular and does not occur in waves, as before acclimatization. For this reason, it is not so noticeable. In addition, there is a shift in sweat production so that there is less on the forehead and more on the trunk and the extremities. The dampness of the face and the sensation of sweating are thus reduced.

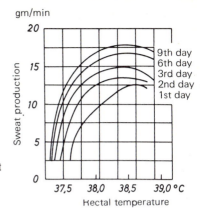

Fig 121   Effect of exercise performed daily in hot climate on sweat production in relation to rectal temperature (according to values from Ladell).

The concentration of salt in sweat, normally about 0.3%, drops to about 0.03% after successful acclimatization. Heat transport through the skin is facilitated. The increased blood flow to the skin is more than compensated for by the increase in total blood volume and by the reduced blood flow to other vascular regions. For this reason, the rise in heart rate for the same work load and at the same climate will be less after acclimatization.

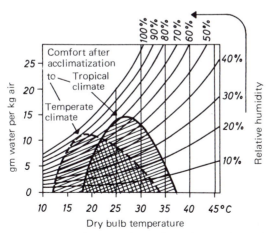

Fig 122   Adaptation of feeling of comfort after acclimatization. Man acclimatized to tropical climate has his optimal feeling of comfort at 26 °C and 80% relative humidity, whereas man acclimatized to temperate climate feels most comfortable at 20 °C and 90% relative humidity.

The sensation of comfort is also improved after acclimatization to heat. Figure 122 shows the comfort zones before and after acclimatization. Man thus represents an organism which can adapt more easily to heat than to cold without using other means. For man to feel comfortable in a cold environment, he must depend on clothes or heating.

4.6 Salt and Water Balance

4.6.1 Distribution of Water in the Organism

With the formation of sweat due to work in the heat, the salt and water balance of the body can become so stressed that this stress can force an interruption of work or even produce death under unfavorable conditions. In view of this danger, it is necessary to discuss the fundamentals of water and salt balance and to analyze their regulation more closely.

The water content of the human body is 50–70% of the total body weight, depending on the proportion of body fat, because fat contains less water than the other components of the organism. Table 13 gives the water content of individual organs of man and their fraction relative to total body weight. Water is found in part within the cells (intracellular fluid) and in part outside the cells (extracellular fluid).

Water is found in blood, in the interstitial fluid and in the cells of the organs. The composition of the fluid in these three sections is basically different. This difference consists primarily in the type of salts dissolved in the fluid. Potassium and phosphate ions are primarily located in the cells; the main ions in the extracellular space are chloride and sodium. The ionic distribution is shown in Figure 123.

The difference in ion distribution is maintained by the active performance of the cells. The concentration of salt in blood and in the interstitial fluid is essentially the same because the capillary wall is permeable to salts. Both fluids have a different protein content, in that

Table 13   Water Content of Individual Organs in Man (A) and Fraction of these Organs Relative to Total Body Weight (B) (from K. J. Ullrich, Wasserhaushalt. In: Kurzgefaßtes Lehrbuch der Physiologie, 4th Ed., ed. by W. D. Keidel, Stuttgart: Thieme, 1975).

	A	B	
Skin	72%	18%	
Muskulature	76%	41%	= 75%
Skeleton	22%	16%	
Heart, Liver, Brain	68–83%	= 10–50%	
Fat	10–30%		

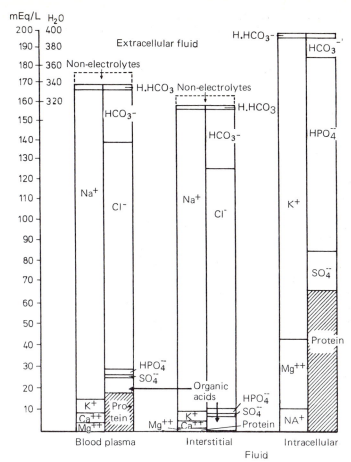

Fig 123   Ion distribution among blood plasma, interstitial and intracellular fluid (according to J. L. Gamble).

the interstitial fluid is predominantly protein free. This is due to the fact that capillaries are practically impermeable to the macromolecular proteins. All three spaces are kept separate from each other by membranes of differing permeability.

Because of these membrane characteristics, water distribution under different conditions can be determined. One can inject indifferent sustances whose concentrations are easy to determine and which distribute themselves uniformly throughout the total body water, e. g., deuterium oxide or antipyrine. The total amount of body water can then be determined according to the following formula:

$$V = g/c, \tag{66}$$

where V is the solvent water volume, g is the amount of the injected substance and c is the concentration of the substance after uniform distribution.

Using the same procedure, the extracellular fluid volume can be determined by taking a substance which cannot penetrate the cell membrane. Examples of such substances are inulin or thiosulfate. Finally, the plasma volume can be determined in the same way by using a substance which cannot penetrate the capillary wall. In this case, primarily albumin labeled with ^{131}I is used.

4.6.2 Movement of Water in the Body

It is not possible to differentiate that water which is transported through the body from that resulting from or necessary for chemical processes; this is so because the molecules are distributed throughout the entire organism. Theoretically, the metabolism of water can be contrasted with the exchange of water. Water metabolism participates in the chemical reactions, whereas water exchange is the path of water through the organism. From the standpoint of exercise physiology, we cannot mention all the details of water movement, e. g., for digestion or the elimination by the kidney of substances bound to urine.

Water also flows in the body in relation to its pressure gradient. The corresponding pressure differences are either hydrodynamically produced pressures, such as those produced by the heart, hydrostatically produced pressures, resulting from the force of gravity, or osmotic pressures, which appear as a result of the number of particles enclosed within a semipermeable membrane. There is also a special type of osmotic pressure called oncotic (colloidal osmotic) pressure. Under normal steady-state conditions, the intracellular and extracellular osmotic pressures are almost the same, whereas the oncotic pressure between the interstitium and the capillaries is different. Pressure differences have various tasks. For example, rinsing of the interstitial space occurs because of the varying pressure differences in the region of the arteries and veins. As we have already seen (p 101), the biggest drop in pressure occurs in the region of the smallest arteries. The oncotic pressure remains essentially constant during the flow of blood from the arteriole to the venule. At the level of the precapillaries, the hydrodynamically-produced pressure predominates and water is filtered from the blood steam into the interstitium. In the region of the venules, the hydrodynamic pressure is less than the oncotic pressure and water is now reabsorbed into the blood stream. In this way, such substances as lactic acid, which appears during hard physical exercise, are eliminated from the extracellular space.

During exercise, a considerable amount of water is displaced from the blood to the muscle due to the appearance of the micromolecular metabolic end products; this evidently produces a rise in hematocrit during exercise.

4.6.3 Disturbances in Water Balance

The distribution of water between the intracellular and extracellular spaces can be disturbed by alterations in either salt or water content. If one drinks low-salt fluids which are hypotonic relative to the blood, the ingested water will be distributed uniformly throughout the intracellular and extracellular fluids. Assuming normal kidney function, this fluid will then be quickly eliminated. If there is a delayed excretion of water (e. g., with kidney diseases), then the cells of the CNS can become so swollen that water intoxication results. The symptoms of this disorder are delirium, disorientation and cramps. Water distribution is also not altered when there is thirst (reduced ingestion). The water contained in the intracellular and extracellular spaces is reduced to the same extent and osmotic pressure increases in both spaces. If the volume of extracellular fluid drops below 70% of normal, then death occurs.

Administration of an isotonic saline solution, which has the same osmotic pressure as blood, causes an increase in only the extracellular space, because there is no change in osmolarity of the blood. Body volume and interstitial fluid both increase to the same degree. In this case as well, salt and water will be excreted by the kidneys, even though the excretion process itself will be much slower. Ingestion of a hypertonic saline solution causes water to leave the cells, until the osmotic pressure in both spaces is equal once more. The opposite situation occurs with a loss of salt. Due to the predominance of the intracellular osmotic pressure in this instance, water enters the cells. During prolonged work in the heat, water intoxication can occur with the symptoms described above. People who work in the heat have been known to lose 6–8 l sweat during a days work. Fluid losses during athletic competition or games in a warm climate are especially large. Inexperienced persons generally try to replace water loss by drinking salt-free fluids. In nonacclimatized persons, this means that up to 25 gm salt may be lost from the extracellular space causing water intoxication; this happens because the blood becomes hypotonic and water is forced into the cells. This water intoxication can take on ominous forms. On the basis of the fundamentals discussed here, the natural conclusion is that one can rapidly eliminate the symptoms by ingesting salt. The salt can be ingested orally, as long as the person is not unconscious. Due to the dangers of aspiration, salt should be administered only parenterally to an unconscious person.

4.6.4 Water Balance and its Regulation

Water balance is adjusted according to the amount of water ingested or produced within the body and the amount of water excreted or consumed by the body. Complete deprivation of food and fluids produces death in a few days; starvation alone with ingestion of fluids can be maintained for up to 60 days. Less fluid is required if glucose is ingested during fluid deficiency because less water is then excreted by the kidneys. This occurs because without food, the body requires protein. The end product of protein metabolism, urea, can be excreted but only up to a certain level of dilution. By giving glucose, protein metabolism is reduced and the loss of water via the kidneys is also restricted. The minimal amount of water to be ingested per day during caloric deprivation is 535 gm.

Water produced by oxidation is governed by the magnitude of the metabolism. The standard value is about 0.16 gm water formed for each kcal metabolism. Thus, 500 ml of water will be produced by oxidation at a daily caloric requirement of 3,100 kcal.

Fluid ingestion is regulated by thirst. The sensation of thirst is governed by the loss of water from the cells. Highly-salted food causes thirst due to the resorption of salt; the extracellular space becomes hypertonic relative to the intracellular space and the cells become dehydrated. Animal experiments have demonstrated that dehydration of a certain group of cells in the middle of the hypothalamus initiates the sensation of thirst.

The excretion of water is regulated by the kidneys, in that more or less water is reabsorbed from the glomerular filtrate. The amount of urine is determined in this manner. The kidney itself, which thus acts as an effector in this control system, is controlled hormonally. The excretion of water is essentially determined by antidiuretic hormone (ADH), which is formed in the supraoptical nuclei of the hypothalamus and transported via the nerve pathway in the supraoptical tract to the neurohypophysis, where it is then stored. A minute amount flows continuously through the bloodstream, inhibiting diuresis. The adequate stimulation for ADH secretion is the elevated osmotic pressure in the hypothalamus or a reduced blood volume, which is monitored by the atrial receptors. A rise in osmotic pressure of less than 1% is sufficient to increase the secretion of ADH and limit the amount of water excreted by the kidneys.

4.7 Physiology of Weightlessness

The problems of weightlessness have been intensively investigated, especially since the beginning of the first space flights. It cannot be an indifferent matter to the reactions and performance capacity of the organism when such an essential factor as the force of gravity is eliminated. The question that many readers will ask is: If this situation occurs so seldom, why is a discussion of it necessary in a book on exercise physiology? This question would certainly be justified if the problem of weightlessness occurred only in space. However, when you sit in a bathtub, you also find yourself in a simulated state of weightlessness. Swimmers today train several hours daily under these conditions. Studies on untrained and endurance-trained persons have shown that both groups react differently to simulated weightlessness of only a few hours.

4.7.1 General Physical and Physiologic Remarks

Force of gravity (= weight) is the force which draws two objects together relative to their mass. Its value is proportional to the product of both masses and inversely proportional to the square of their distance from each other. This means that the force of gravity approaches zero as the distance between the objects approaches infinity. In principle, the force of gravity is present everywhere.

According to the law "actio = reactio" (Newton), the force of two opposing objects is the same.

We will consider the situation where one of the two masses is the earth. If the earth and another object are in direct or indirect contact, they exert a pressure on one another at the surface of that contact. Depending on the solidity of the object, this can produce deformation and inhibit acceleration. On the other hand, if there is a gap between them, then the earth and the object will be accelerated toward each other: Force = mass · acceleration.

Because the mass of the body is small relative to that of the earth, we are interested only in the acceleration of the body in the direction of the earth's center of gravity; this is called acceleration due to gravity or due to a fall. It has a value of about 9.81 m/sec^2 at the level of the earth's surface; the opposing acceleration of the earth can be disregarded. The state in which a body can act without the influence of gravity and where no pressure differences exist between the different parts of the body is called weightlessness. To investigate weightlessness and its effects on man, expensive experiments are needed, because the subjects must be transported into space. Due to limitations in weight and space, the extent to which physiologic studies can be performed in a space laboratory is limited. For these reasons, man has endeavored

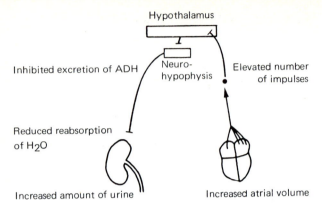

Fig 124   Summary of Gauer-Henry reflex. From Schneider, M.: Einführung in die Physiologie des Menschen. (Berlin: Springer, 1971).

for some time to simulate weightlessness in order to imitate some of the effects and to make them comparable to those found during experiments in space. Immersion in thermoindifferent water is an especially good alternative.

Although the principles of volume and osmolarity regulation have already been discussed on page 120, they should be discussed further at this point, because all of the main disturbances initiated by weightlessness are related to them.

Figures 124 and 70 show the basic flow of information in the control system regulating blood volume and the osmotic pressure in blood. Receptor areas are found primarily in the left atrium of the heart and continuously monitor the dilation of the atrium. It should be made clear why this important blood depot of the circulatory system is found in this region, i. e., because the blood vessels of the lungs, as well as those of the atria of the heart, have the greatest extensibility in the circulatory system. If fluid is introduced into the circulation due to eating or drinking, then intravascular volume will increase, especially at those sites in the circulatory system which are the most extensible. As a result, the atrial receptors will be activated. Their information is transmitted neurally to the hypothalamus and to the hypophysis. From here, the information continues hormonally, i. e., the release of ADH is inhibited. Acting via the bloodstream, ADH controls the excretion of urine. Therefore, a reduction in ADH always means an increased excretion of urine from the kidneys into the bladder.

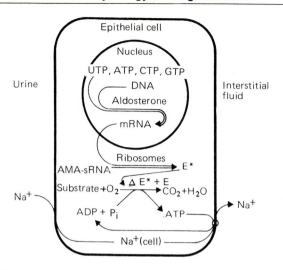

Fig 125   Possible operating mechanism of aldosterone in the kidney. UTP, ATP, CTP, GTP are purine- and pyrimidine-triphosphate derivatives; mRNA, messenger RNA; AMA-sRNA, amino acid-sRNA compound; E, enzyme; E *, newly synthesized enzyme; P_i, inorganic phosphate (according to Edelmann, Bogoroch and Porter in: Ganong, W. F.: Medizinische Physiologie (Berlin: Springer, 1971).

In this control system, the excretion of water from the blood adapts to the filling of the circulatory system, so that blood volume is maintained at a fairly constant level. A second input continuously monitors osmotic pressure of the blood; this is primarily determined by the concentration of salt in the plasma. Because concentration is related to the amount of the substance per unit of solvent water, it may be adjusted either by excretion of salt or by excretion of water. This first control system adjusts loss of water via the same ADH mechanism. Its adjustment time is relatively short, i. e., the adjustment process is activated within a few minutes.

The second control loop in this interconnected control system, whose manipulated hormone is aldosterone, is adjusted by the same variables, i. e., volume and sodium concentration. Aldosterone causes a reabsorption of sodium in the renal tubules. The extension volume receptors lie primarily in the right atrium. Although ADH secretion is stimulated via the left atrium and develops its full effect within 1–2 minutes, the effect of aldosterone begins and ends about one hour later. The mechanism for the effect of aldosterone can be presented (Fig 125) in the following manner. There are number of purine- and pyrimidine-triphosphate derivatives located in the cell nucleus.

Together with DNA and aldosterone, they form messenger RNA, permitting the neogenesis of an enzyme, which causes the restitution of ATP from oxidative metabolism. This process probably also is the reason for the slow response time, i. e., protein synthesis for the formation of enzymes occurs slowly.

Adenosine triphosphate has various tasks to fulfill in the body (p 44). It drives the sodium-potassium pump of the cells, assuring the constancy of the membrane potential. In addition, it provides the energy after a contraction to pump calcium from the sarcoplasm back into the lateral sacs. Of greatest importance is the fact that ATP is the immediate source of energy for muscle contraction. If its restitution is impeded, then all of the ATP-dependent processes will also progress more slowly. In the cell shown in Figure 125, for example, the active transport of reabsorbed sodium will occur more slowly, because ATP provides the needed energy. Thus, more sodium will be excreted.

4.7.2 Hemodynamic Effects of Weightlessness

Depending on the extensibility of his veins, a man standing upright will normally have 400–800 ml of blood in the region of his lower extremities; this is caused by hydrostatic pressure. As a result of the force of gravity, blood accumulates in those veins which represent the lowest point at any given time. Because the column of blood is shorter while lying, the volume in the leg veins is less and the intrathoracic volume is greater. Bed rest therefore causes some of the same reactions found during weightlessness, although in a milder form. There is a displacement of blood during the transition from standing to lying, where the surplus blood is deposited in the "central blood depot," i. e., in the lungs and in the atria of the heart.

Under conditions of weightlessness, hydrostatic pressure reaches a value of zero. A part of the blood normally found in the dependent portions of the body fills up the central blood depot and then stimulates the atrial receptors in the same way as occurs under physiologic conditions due to ingestion of fluids. A rough calculation will clarify the strength of this stimulus.

It is estimated that about 30% of the total blood volume is found in the central depot at rest. Therefore, if about 600 ml fluid displaces itself into the central blood depot, this will have the same effect as if 1.8 l fluid was ingested. Due to weightlessness, the system for regulating blood volume receives false information. Thus, stimulation of the sensor is not produced by the increased blood volume, but by its altered distribution. Simulation of weightlessness is based on the fact that one can compensate for hydrostatic pressure by buoyancy in the water. The hydrostatic pressure exerted by water on the superficial veins empties these veins, causing a modification in blood distribution

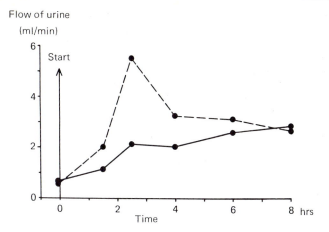

Fig 126 Pattern of urine excretion during immersion of untrained (dashed line) and trained (solid line) persons.

similar to that found in true weightlessness. This procedure is called immersion. Because the body is easily warmed or cooled, due to the different heat conduction properties of water versus air, the indifferent temperature of 35 C must be exactly maintained, especially as only prolonged studies are of interest.

4.7.3 Physiologic Effects of Simulated Weightlessness

4.7.3.1 Excretion of Urine

In each of the following presentations, the mean values of 4 untrained and 4 endurance-trained subjects are compared. The curves of the trained subjects are represented by solid lines and those of the untrained subjects by dashed lines. The vertical line denotes the onset of immersion. Figure 126 shows the pattern of urine excretion for both groups during 8 hours of immersion. Although the flow of urine from the trained subjects rises continuously from beginning to end, the reaction of the untrained subjects exhibits a marked overshoot.

Despite the elevated excretion of urine, there is little change in osmotic pressure of the plasma during immersion. At first, this may seem surprising, since an increased excretion of water due to a reduced secretion of ADH usually produces a rise in osmotic pressure. However, in addition to the augmented excretion of water, the excretion of

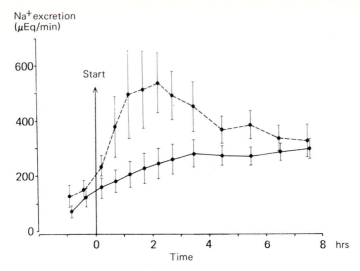

Fig 127   Pattern of sodium excretion during immersion of untrained (dashed line) and trained (solid line) persons. (Compare with Figure 130, in which control experiments in lying position out of the water are shown).

sodium is also increased (Fig 127). Because chloride and sodium are bound together, more chloride is also excreted. Thus, it is understandable why the osmotic pressure of both untrained and trained subjects remains constant. The control diagram in Figure 70 indicates that osmotic pressure influences water excretion negatively. The effective water loss is the result of two opposing influences. If trained subjects react with a slower excretion of water while maintaining a fairly constant osmotic pressure, this can mean only that their volume control system for the excretion of water reacts less sensitively. This could be due to a change in the geometry of the heart, the response to extension of the myocardial fibers or the circulatory center. A highly-trained endurance athlete has about 10–15% more total blood volume at the same body weight. If his control system had always maintained his blood volume at a fixed set point from the onset of training, then he would never have been able to reach a higher blood volume. It is possible to conclude that either the set point was adjusted or the sensor itself become less sensitive during training.

The loss of water from the blood means that there will be a rise in concentration of all those substance for which the capillary wall is impermeable. This fact can be clearly seen in the hematocrit value (Fig 128) because there is a definite rise in both groups. The hematocrit is

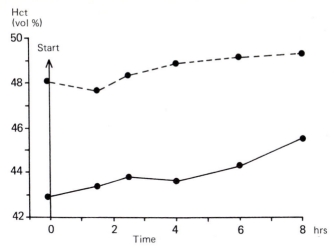

Fig 128 Changes in hematocrit during immersion of untrained (dashed line) and trained (solid line) persons.

defined as the fraction of blood cells in the total blood fluid. The difference in the initial level for trained and untrained subjects corresponds to the known fact that endurance-trained athletes have a lower hemoglobin concentration (p 291). There are no systematic modifications in the average values for protein concentration, especially the albumin fraction.

4.7.3.2 Aldosterone and Electrolyte Concentration and Excretion during Immersion

The method of determining aldosterone concentration in blood is still expensive. There is an indirect method available, in which the aldosterone excreted into the urine is related to the glomerular filtration rate of the kidneys. This can be considered a relative measure of the plasma concentration. In Figure 129 it can be seen that there is a drop in measured aldosterone concentration at the beginning of immersion. This is the basis for the following events.

As would be expected with a reduced excretion of aldosterone, there is a marked rise in sodium excretion (Fig 127) during immersion. Similar to their responses with urine excretion, trained subjects have a diminished reaction; this suggests a lower inhibition of aldosterone secretion. Because aldosterone, in addition to other influences, is controlled by the volume receptors in the right atrium, this also suggests a diminution in the ability to regulate volume after prolonged

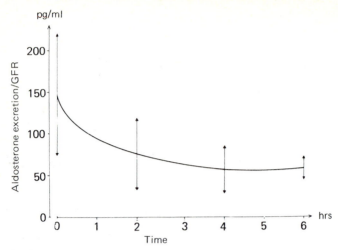

Fig 129   Excretion of aldosterone per ml glomerular filtration rate before and during 6-hour immersion. pg = 10^{-12} gm (according to Skipka).

endurance training. The sodium concentration in plasma is not systematically altered. This is explained by the fact that there is a proportionate rise in the excretion of both water and sodium.
Of special interest is the response of inorganic phosphate (Fig 131). The plasma concentration of inorganic phosphate increases in both

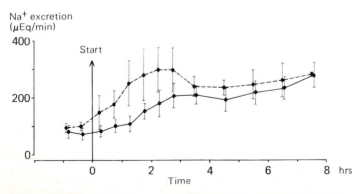

Fig 130   Pattern of sodium excretion in untrained (dashed line) and trained (solid line) persons while lying down out of water. (Compare with Figure 127, in which corresponding values during immersion are shown), (according to Skipka).

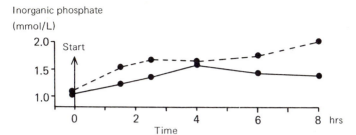

Fig 131 Concentration of inorganic phosphate in plasma before and during immersion of untrained (dashed line) and endurance-trained (solid line) persons.

groups. After the onset of aldosterone inhibition, there is a marked rise in excretion into the urine in both groups. This finding evidently confirms the hypothesis of enzymatic induction for the total body as well. When less ATP is resynthesized from ADP and inorganic phosphate (due to the fact that the aerobic energy supply is diminished), more and more inorganic phosphate must appear in the cells and then diffuse into the plasma.

This mechanism, which up to this point has been postulated only for the reabsorption of sodium, gains new dimensions when it is applied to maximal performance capacity before and after weightlessness. In this case, ATP has a particularly important role because it transmits the energy directly to the contractile proteins (p 15).

4.7.4 Performance Capacity and Weightlessness

Performance capacity has two important levels: maximal performance capacity, which is especially important for sport, and endurance capacity, which is associated more with occupational work. The upper limit of endurance capacity is related to the threshold for prolonged work (also called the endurance threshold) and is characterized by the fact that below this endurance threshold, oxygen requirements of muscle and the transport of oxygen to its mitochondria are balanced. For this reason, there is only the initial energy debt, which remains constant over the total period of work (details, p 64).

To compare maximal performance capacity before and after weightlessness of 6 hours, the maximal oxygen intake can be ascertained during exhausting exercise on a bicycle ergometer. For healthy persons, this is primarily limited by the performance capacity of the heart, capillarization and the enzyme activity of the muscles. The second useful test, Müller's performance heart-rate index (LPI), yields information about the threshold for prolonged work. It provides data at

Table 14   Changes in Maximal Oxygen Intake due to 6-Hour Immersion in Thermoindifferent Water

	Before Immersion l/min (STPD)	After Immersion l/min (STPD)	Changes in % of Initial Value
Untrained Mean and SD	3.81 ± 0.44	3.33 ± 0.37	10.03 ± 7.05
Trained Mean and SD	4.77 ± 0.48	3.9 ± 0.38	19.34 ± 8.79

submaximal work loads about the rise in the number of heart beats per minute needed for an elevation in work load of 1 kpm/second.

Table 14 shows the results of experiments before and after immersion in thermoindifferent water for a period of 6 hours. In the left-hand column, the maximal oxygen intake (l/minute, STPD) is given. The mean value of the untrained subjects before immersion is clearly less than that of the trained group; this is an indication of the higher performance capacity in the trained men. After immersion, maximal oxygen intake is somewhat lower in the untrained subjects. On the average, there was a drop of about 10% from the initial value. With the trained men, maximal O_2 intake of some subjects decreased up to 25–30%, with an average decrease of 20%. An endurance-trained man after simulated weightlessness reacts somewhat like an untrained man without simulated weightlessness.

Müller's performance heart-rate index exhibits much the same response (Table 15). Before immersion, the untrained have a higher average value than the trained subjects; the index is higher when performance capacity is lower (p 246). After immersion, there was an increase of about 28% in the trained men, but only about 14% in the untrained men. This test also shows that both groups have a reduced endurance capacity, even though the drop is much greater in the trained group.

Table 15   Changes in Performance-Heart Rate Index of E. A. Müller due to a 6-Hour Immersion in Thermoindifferent Water

	Before Immersion	After Immersion	Changes in % of Initial Value
Untrained Mean and SD	3.28 ± 0.72	3.75 ± 1.11	13.9 ± 20.1
Trained Mean and SD	2.65 ± 0.67	3.21 ± 0.68	28.3 ± 25.9

Both maximal performance capacity and endurance capacity decrease and the reduction is much greater in trained men. It could be hypothesized that this is related to a disturbance of ATP resynthesis, due to a reduction in aldosterone activity.

4.7.5 Orthostatic Tolerance and Weightlessness

Orthostatic tolerance is the ability to regulate blood pressure during transition from the lying to the upright position. A volume of blood (400–800 ml) is displaced to the lower extremities from the central blood depot. As a result, there is less filling of the heart, producing a drop in stroke volume. Initially, there is no alteration in heart rate and cardiac output becomes smaller. Because peripheral resistance remains constant at first, mean arterial pressure decreases. At this point, control via the baroreceptors begins, causing an increase in sympathetic tone to the heart, as well as to the peripheral vessels. The primary result is a rise in diastolic pressure, partially compensating for the initial drop in mean arterial pressure. In addition, heart rate and stroke volume increase as much as possible, given the available reserves in the central blood depot, and there is full compensation of the decreased cardiac output. If the control fails to operate, there will be a collapse, primarily due to the fact that blood pressure cannot be maintained at a level sufficient to perfuse the brain. This results in a characteristic symptom called vasovagal syncope.

Typical data found in such experiments can be seen in Figures 132 and 133. Consider first, graphs A at the top of these figures. These show the response of blood pressure variables and heart rate, as a function of body position, in trained and untrained subjects. Both groups show basically the same responses, even though they start from different levels.

After an immersion of 6–8 hours (A below), untrained men have a narrowing of blood pressure amplitude and a sharp rise in heart rate; this is an obvious indication that sympathetic tone must have been markedly activated to balance the mismatch between the desired and actual values. A short time after assuming the upright position, trained subjects (A below, Figure 133) regularly showed symptoms of vaso-vagal syncope with collapse and the experiment had to be stopped.

It can be clearly seen that the orthostatic tolerance of all subjects was diminished and that the compensatory mechanisms of the untrained evidently were still adequate to avoid collapse, whereas the trained subjects obviously surpassed their range of control.

The cause of the orthostatic weakness in trained men probably is due to a more even control characteristic for the regulation of their blood pressure (Section 2.6.4). In addition, hypoaldosteronemia could have a role, as patients with hypoaldosteronemia due to disease also exhibit

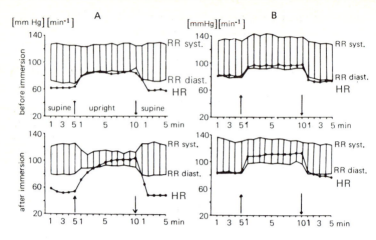

Fig 132 Circulatory reactions of untrained subjects during passive changes in body position from 0 to 90 degrees before and after a 6- to 8-hour immersion. A, immersion without work (4 subjects); B, immersion with intermittent work (9 subjects); RR, arterial blood pressure values obtained using the method of Riva-Rocci; HR, heart rate.

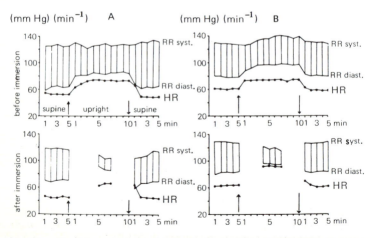

Fig 133 Circulatory reactions of trained subjects during tilt-table experiments (same as in Figure 132) before and after 6- to 8-hour immersion. A, immersion without work; B, immersion with intermittent work; RR, arterial blood pressure values obtained using the method of Riva-Rocci; HR, heart rate. After immersion, entire experiment could not be completed due to collapse.

poorer orthostatic tolerance. It is also possible that the effect of norepinephrine on the vessels is diminished.

The graphs under B show for both groups that orthostatic intolerance is less if the subject is intermittently exercised during immersion. In the present investigations, subjects were exercised to exhaustion for 3 minutes at the beginning of every hour. As a result of this work, micromolecular lactic acid accumulated in the muscles; this evidently produced water displacement due to osmotic pressure, with a resulting reduction in central blood volume, and thus a lessening of those effects normally seen with immersion. It appears that this is the reason why swimmers do not exhibit orthostatic dysregulation, even after prolonged training in water.

4.7.6 Effects of Actual Weightlessness

For reasons given in the introduction to this chapter, there is little material available in the published literature about the comparison of true to simulated weightlessness. Nevertheless, the relationship between heart rate and defined work loads before and after a sojourn in space has been determined in detail, especially during the space programs of Apollo 7–11 (Rummel et al. 1973) and Apollo 14–17 (Rummel et al. 1975). It was shown that heart rate at the same work loads was markedly higher during the first hours after return to gravity

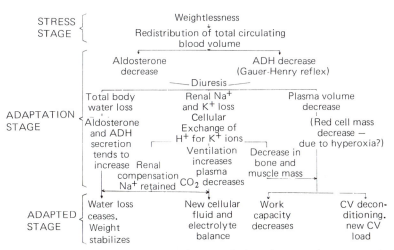

Fig 134   Diagram of hypothesized course of adaptation to prolonged weightlessness (according to Berry).

and that this was similar to the effects already discussed on page 210 in relation to the performance heart-rate index. After several days, these changes were less noticeable.

There were also demonstrable changes in orthostatic tolerance after space flights (Hoffler et al. 1974).

Even under conditions of weightlessness, regular physical exercise produces a training effect on maximal O_2 intake, as was found in the studies of Space-Lab (Savin et al. 1975). This is not necessarily surprising because swimmers, who evidently train under simulated conditions of weightlessness, show a marked improvement due to training, even though it is somewhat less than that found after training out of water.

Prolonged effects can be studied especially well in true weightlessness. According to the concept of Berry (1974), there are certain adaptation phenomena during the course of weightlessness which do not affect all systems in the same way. The course of adaptation to prolonged weightlessness is diagrammed in Figure 134. The stress stage has already been studied in detail. Little is known today about the adaptation stage; it is possible that more exact evaluations of the Space-Lab studies will produce more conclusions.

4.8 Physiology of Diving and Swimming

4.8.1 General Physical and Physiologic Remarks about Diving

If we grasp the fact that the total column of air over the earth exerts the same pressure as a column of water 10 m high, it is understandable how dangerous diving at seemingly harmless depths can be, if the diver does not pay attention to the pertinent physical and physiologic laws. The primary dangers occur because pressure differences can appear within the body which are related to the different physical responses of gases, fluids and solid objects under pressure. Furthermore, the modified blood gas pressures can produce dangerous disturbances.

4.8.1.1 Boyle-Mariotte Law

Pressure is defined as the ratio of force to surface. The unit is 1 kp/cm^2 equals 1 atmosphere (atm) equals 760 mm Hg equals 1 bar. The absolute pressure in atmospheres is called the ata (absolute atmosphere). The normal atmosphere of 760 mm Hg is exactly the same as 1 ata. If one takes as the comparison pressure that of the actual atmosphere, then one talks about atmospheric excess pressure. An automobile tire which is pumped up to the prescribed pressure of 2 kp/cm^2

would thus have 2 units atmospheric excess pressure but 3 ata because the comparison pressure in this instance is the actual air pressure.

In contrast to fluids, gases are compressible. An ideal gas obeys the Boyle-Mariotte Law:

$$P \cdot V = \text{constant},$$

where P is pressure and V is volume.

If a rubber bulb is filled with 1 l air at the normal atmospheric pressure of 1 ata at sea level, there will be double the pressure at a water depth of 10 m (2 ata). At the same time, the volume will be compressed to 0.5 l. At depths of 20 m and 30 m, there will be only one-third and one-fourth of the initial volume, respectively.

4.8.1.2 Partial Pressure of Gases

The partial pressure of a gas is the fractional pressure exerted by that gas; this is proportional to its concentration relative to total pressure. In a dry mixture of 30% gas A and 70% gas B at 1 ata (160 mm Hg), the partial pressure of gas A is 228 mm Hg and that of gas B is 532 mm Hg. Under conditions of 2 ata, the partial pressures will be doubled to 456 and 1,064 mm Hg, respectively. The same conditions are also valid for respiratory gases, with the addition of water vapor found in the air. Water vapor has the characteristic of exerting a pressure which is dependent only on the degree of water vapor saturation and on the temperature. To determine the partial pressure of respiratory gases in the lungs, complete saturation with water vapor and a temperature of 37 C, are generally assumed, thus obtaining a water vapor pressure of 47 mm Hg. The total pressure must then be reduced by this amount in order to convert it to "dry conditions" (STPD)*. Water vapor pressure, as a function of temperature, can be obtained from tables, e. g., *Documenta Geigy*.

4.8.1.3 Solubility of Gases in Fluids

The solubility of gases in fluids varies from one gas to another. The concentration of an ideal gas in this case is proportional to its partial pressure:

The proportionality factor α is the physical solubility or the solubility coefficient with the dimension, amount of gas per volume per partial pressure. Within the field of medicine, the Bunsen solubility coefficient is often used with the unit: ml gas (STPD) per ml fluid per

---

* Standard Temperature Pressure Dry equals 760 mm Hg, 0 C, dry

760 mm Hg. Solubility is dependent on the nature of the gas, on the dissolving fluid and on the temperature.

4.8.2 Apneic Diving

4.8.2.1 Pressure Relationships during Deep Diving

Apneic diving (apnea means without breathing) is defined as diving under the water without technical aids. The problems associated with this simplest form will now be made clearer, which will help later when we discuss diving with an apparatus. The main problem is that within the body, air-filled cavities must adapt to volume changes due to variable pressures, even though the pressure differences within the body occur which produce trauma (injury); they are therefore classified under the collective term of "barotrauma."

The largest air-filled cavity in the human body is the lung. Depending on the depth for which he is striving, the apneic diver fills his lungs with air on the surface up to 75–100% of its total capacity. If he dives to a depth of 10 m, placing his lungs under an external pressure of 2 ata, then the position of his thorax goes from an almost maximal inspiratory position to a mid-expiratory position without any air escaping. As a result, the pressure in the lungs is doubled and the volume is halved. The content of the lungs has adapted to the surrounding water pressure by means of compression (Fig 135).

The mean arterial pressure in healthy persons of about 100 mm Hg equals 0.13 units of excess atmospheric pressure. This means that under normal conditions, it is 100 mm Hg greater than that of the environmental air pressure. How high is it when at a water depth of 10 m? To answer this question, we must assume that the water pressure is spread evenly over all tissues and, at least from a physical standpoint, that the pressure differences will not be altered. Mean blood pressure, which on the surface has an absolute pressure of 760 mm Hg + 100 mm Hg = 860 mm Hg = 1.13 ata, accordingly has a pressure of 2 × 760 + 100 = 1,620 mm Hg = 2.13 ata at a depth of 10 m. In the same way, it is possible to calculate the pressure in the pulmonary circulation, which has a pressure of 5–10 mm Hg compared to that of the environment. Under normal atmospheric conditions, the pressure difference in the inner cavity of the lungs is about 0–12 mm Hg because the pressure in the inner cavity varies ±1–2 mm Hg between inspiration and expiration. At a water depth of 10 m, the pressure difference does not change, because the internal pressure of the lungs, as well as the pressure in the pulmonary circulation, are both increased to the same extent by compression.

The maximal depth of apneic diving is essentially determined by the capacity of the lungs. There are some considerations which should be

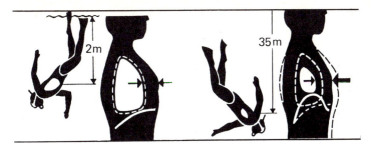

Fig 135   Schematic representation of relationship between lung volume and depth of diving. With increasing depth of diving, air becomes compressed, such that thorax with same amount of air goes from inspiratory position to expiratory position. From Ehm, O. F. and Seemann, K.: Sicher Tauchen. Rüschlikon-Zürich: A. Müller 1965).

Fig 136
Schematic representation of pattern of blood volume, residual (Res Vol) and vital capacity (VC) with changes in depth of diving. A, pattern at the surface. Blood is under normal atmospheric pressure and ratio of residual volume to vital capacity remains about 1 : 4. B, water depth of 10 m. Blood assumes same volume, even when pressure is double that found under atmospheric conditions because fluid is not compressible. According to Boyle-Mariotte Law (P · V = constant), vital capacity + residual volume are compressed to one-half their original volume. There now exists no pressure gradient between blood and lungs. C, greatest diving depth man can reach without having pressure gradient. Thorax is in its maximal expiratory position

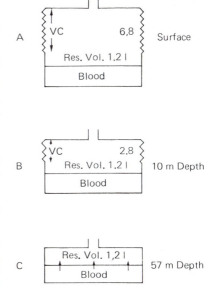

and air volume cannot be further compressed. For this reason, lung contents cannot become equilibrated with pressure in blood if diver goes still deeper. As a result, pressure differential with inner cavity of lungs exists and might produce barotrauma.

made clear in Figure 136. Total capacity of the lungs can be divided into two functional volumes: vital capacity, i. e., the portion of air which can be voluntarily expired after a maximal inspiration, and residual volume, which is the amount of air still remaining in the lungs afterward (p 146). This latter volume is primarily due to the bony thoracic cavity, which cannot be arbitrarily emptied by pressure.

The decisive factor for maximal diving depth is the degree to which the pressure between the blood stream and pulmonary cavity can be adapted via volume reduction. Using data from a top athlete whose vital capacity is 6.8 l (normal persons 4–5 l) and who thus has a total capacity of about 8 l, a maximal diving depth of 57 m can be calculated, without having the problem of a pressure difference. Because blood will evidently still be shunted into the thoracic cavity, a person can reach even greater depths without injury, as long as he stays only a short time at these depths. The world record as of Autumn 1975 was 92 m (Jacques Mayol). Among other air-filled cavities, the most important ones for diving are the middle ear and the sinuses. The middle ear can be functionally connected with the pulmonary cavity via the eustachian tube and the oral cavity, as long as the ducts are free; they can be obstructed, especially by catarrh associated with inflammation of the mucous membrane. The tube is normally opened by swallowing, thus equalizing the pressure. The sinus cavities can also be connected with the pulmonary cavity via the windpipe.

4.8.2.2 Circulatory Reflexes during Apneic Diving

Blood gas pressures are considered important in this discussion because only when there is sufficient oxygen pressure can the brain, liver and heart remain functional. Other tissues can contract a greater or lesser oxygen debt. Diving animals (whales, seals, ducks) have developed special mechanisms of adaptation, which are also seen in man, although in a much less pronounced form. The most important of these will be mentioned here because they demonstrate clearly the problems involved. They allow the animals to dive for much longer periods than an apneic human diver can. Experienced divers have a maximal breathholding time of 3–4 minutes, whereas the whale can remain under water for up to 60 minutes. Most diving animals breathe out maximally before diving, so that for all practical purposes, only the dead space is filled with air. They essentially have no residual volume left because the thorax is not rigid. As a result, during long periods of diving which would be impossible for man without a breathing apparatus, no nitrogen will be dissolved into the blood and thus, no caisson disease (p 224) will occur after a rapid rise to the surface.

This reflex allows diving animals to concentrate their entire circulation on blood flow to the brain and heart, i. e., they can contract the smallest arteries in other parts of the body so that blood flow there will be completely cut off. Thus, the oxygen present in the blood is used for the aerobic basal metabolism of the brain and heart. Compared to man, the musculature of these animals has a higher myoglobin content and a greater O_2 reserve. In addition, they can contract a larger O_2 debt. Heart rate is very much reduced during diving.

Diving bradycardia continues in man, even during or after heavy physical work. The reflex overrides all known mechanisms of circula-

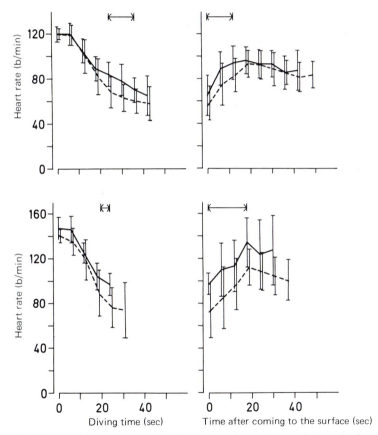

Fig 137   Graphic representation of reduction in heart rate during simple breathholding and during diving. Graphs on right show changes in heart rate after completion of apnea. Area of significant difference is marked by arrows (dashed line, diving; solid line, simple breathholding). From Stegemann, J. and Tibes, U.: Pflügers Arch. 308 : 16, 1969.

tory drive. A part of the effect can even be seen during simple inspiratory breathholding.

Figure 137 shows results from studies where subjects first swam with varying intensities against an elastic band. During the exertions against the elastic band, heart rate increased over the 2–3 minute period of swimming. When the subject's heart rate reached the desired value, he remained at rest for about 10 seconds and then went about 1 m under the water by pulling on an object lying on the bottom of the pool. After a latent period of about 10 seconds, the diving maneuver itself produced an abrupt fall in heart rate to values which were mostly under the resting value. Immediately after rising to the surface, heart rate rose once more within 20 seconds but still did not reach the value seen before diving. It then slowly returned to the resting value in the usual manner (*dashed line*). Results from comparative experiments, in which the subjects held their breath while lying in water, are shown (*solid line*). The heart rate reaches its minimal value of 50–60 beats per minute after about 30–40 seconds of diving.

This bradycardia is considered to be a vagal effect because it can be suppressed by giving atropine. However, diving bradycardia is not just the effect of a single influence but the sum of numerous factors. The greater intrathoracic blood volume in water caused by relative weightlessness and by hydrostatic pressure has just as important a role as the pressure relationships in the pulmonary circulation and the altered conditions of venous return produced by water. In addition, oxygen deprivation must be included in this discussion. Whether man also has receptors in the region of the trigeminal nerve, as was found in the diving reflex of ducks, is not yet known.

To understand the mechanism, the effects of a Valsalva maneuver done out of water must be understood. If an individual breathes out against a given pressure, blood pressure amplitude is reduced and heart rate rises markedly. This effect indicates that the large veins are compressed by the elevated intrathoracic pressure, greatly reducing the flow of blood to the right side of the heart. Only a small amount of blood will be pumped from the right side of the heart to the pulmonary circulation; the left side of the heart acutely reduces the intrathoracic blood volume due to its pumping effect. After a few beats, it will be insufficiently filled and pressure in the "Windkessel" arteries drops off. This results in a release of inhibition of sympathetic tone. At the same time, the reduction in blood pressure amplitude also produces a release of inhibition of the sympathetic nerve. Both influences produce a rise in heart rate.

The situation in water is different. Due to hydrostatic pressure, most of the superficial veins at least are emptied and the central blood volume is increased. The Valsalva maneuver produces different reactions, depending on the depth of the upright body in water (Fig 138).

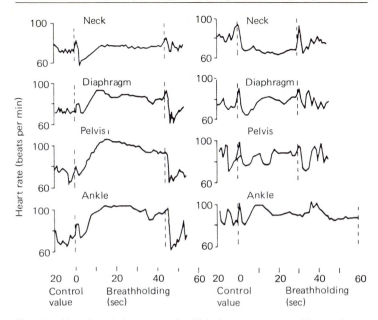

Fig 138 Alterations in heart rate after Valsalva maneuver at different depths of immersion of the body in water. From Craig, A. B.: J. Appl. Physiol. 18 : 854, 1963.

Although heart rate increases while performing a Valsalva maneuver in air, it drops when the body is immersed to the navel. Therefore, the assumption is that the greater venous blood supply now causes a rise in blood pressure and an enlarged amplitude. This, in turn, inhibits the sympathetic nerve via the baroreceptors, causing a reduction in heart rate. If this interpretation is correct, then the inhibition of sympathetic tone should produce a drop in heart rate, along with a reduction in peripheral resistance. On the other hand, systolic pressure usually remains constant during diving, whereas diastolic pressure is slightly elevated. This essentially constant blood pressure with definite brady-cardia can only be explained by a combined effect of a rise in sympathetic tone with marked vasoconstriction in the peripheral circulation and an isolated rise in vagus tone.

A good diver usually fills his lungs to about 80% of maximal inspiratory volume before the dive and closes his glottis to prevent the loss of this volume. This produces a slight excess pressure of several mm Hg in the lungs, which increases the transmural pressure of the total pulmonary circulation, as well as that of the vena cava. If the diver were in air, a pressure of this magnitude would produce two mechanical effects. The large veins would be compressed, thus reducing the inflow to the right side of the heart. At the same time, resistance in the pulmonary circulation would rise due to the elevated transmural pressure. According to the equilibrium curves seen on page 108, the right side of the heart under constant innervation conditions can overcome the elevated resistance by developing a higher pressure only if it increases its residual volume. However, the limited influx prohibits this possibility.

During immersion in water, intrathoracic blood volume and residual volume of the right side of the heart are markedly higher. In addition, pressure in the intrathoracic veins is slightly higher due to the displacement of blood. Inspiration before the dive now produces a further stretching of the heart, aorta and pulmonary arteries via a reduction in intrathoracic pressure. Experiments with cats have demonstrated that stretching of the right side of the heart, aorta and pulmonary arteries causes marked bradycardia. Apparently there are protective reflexes for the right side of the heart coming from this region, similar to those for the left side of the heart described by the Bezold-Jarisch reflex. Bradycardia is apparently initiated only when a strong stretch stimulus is applied to the right side of the heart, aorta and pulmonary arteries through an inspiration combined with immersion. Thus, diving bradycardia is caused by two effects: a general elevation of sympathetic tone, which (along with other causes) can be produced by a lack of oxygen in the periphery, and an isolated rise in vagal tone, which is induced via baroreceptors in the region of the truncus arteriosus.

4.8.2.3 Blood Gas Pressures during Diving

Figure 103 indicates the time course of the drop in O_2 pressure in alveoli when the breath is held. Loss of consciousness occurs when alveolar O_2 pressure goes below about 30 mm Hg and arterial O_2 pressure goes below 20 mm Hg.

Oxygen supply to brain tissue depends on the pressure difference between capillary blood and the mitochondria, the site of aerobic metabolism. During deep diving (at a depth of 10 m), total pressure and also the partial pressure of oxygen (P_{O_2}) in the lungs is doubled. Because there is already a lower P_{O_2} in the mitochondria than in the capillaries under normal conditions, the pressure difference during diving will be greater. The supply to the tissues will generally be improved under high pressure so that the oxygen present will be better utilized. During the ascent to the surface, there is a rapid drop in total pressure and in P_{O_2} in the lungs, so much so that it is possible to reverse the direction of diffusion. The ascent phase is especially dangerous and as a result of oxygen lack, such difficulties as disorientation or loss of consciousness can occur.

In distance swimming underwater, where water pressure generally has a minor role, one should pay attention to the fact that loss of consciousness can appear suddenly if the oxygen pressure in arterial blood goes below the critical value. The drop in P_{O_2} is associated with O_2 utilization.

The diver is usually forewarned by the appearance of a strong need to breathe; this is associated with the respiratory drive and represents an important safety factor. It would be dangerous to suppress this respiratory drive and its primary components of CO_2 and hydrogen ion concentration by voluntarily hyperventilating before the dive. With hyperventilation, large amounts of CO_2 are blown off so that the lack of oxygen can become effective without any forewarning. This is especially true for endurance-trained individuals because the sensitivity of their respiratory center to CO_2 pressure is reduced.

4.8.3 Diving with an Apparatus

In apneic diving, the limited diving time can be a disadvantage. Therefore, there are a number of devices being used, especially in leisure sports. The simplest of these is a snorkel, which is a curved tube with a mouthpiece on one end and a free opening at the other. The snorkel allows a maximal diving depth of 35 cm while still being able to breathe. Using a longer or lengthened snorkel can be dangerous and can produce barotrauma of the lungs for the following reasons.

Assuming that a person dives with a snorkel to a depth of 1 m, then the absolute blood pressure in the pulmonary vessels increases to about 76 mm Hg. On page 216, it was shown that the pressure in the thoracic cavity has a compensatory rise during apneic diving due to the compression of air (Boyle-Mariotte law). This is not possible during snorkeling because the thoracic cavity is connected to the external atmosphere via the snorkel. The pressure difference between the thoracic cavity and the pulmonary circulation thus increases to 76 mm Hg at a depth of 1 m. The lungs then become so filled with blood that respiratory difficulties and destruction of the blood vessels can result. This condition is called barotrauma of the lungs.

4.8.3.1 Diving with Automatic Lungs

An automatic lung is an apparatus which permits the necessary amount of air to enter the lungs while underwater. There are generally two systems. In the open system, air is breathed in from tanks of pressurized air and expired air is breathed out into the surrounding water. In closed system, carbon dioxide is absorbed and the oxygen used is replenished.

Each apparatus must be constructed to adapt the pressure of the inspired air to that of the surrounding water via compression; in this way, no pressure difference between the thoracic cavity and the blood stream occurs. The pressure-adapting mechanism must react rapidly, so that no pressure differences appear, even during transition phases.

4.8.3.2 Special Dangers during Apparatus Diving

Although the apparatus allows longer and deeper dives, two especially important dangers, caisson disease and oxygen poisoning, must be remembered. Caisson disease (a caisson is a diver's bell) refers to the effect of a sudden drop in pressure and is also called decompression disease. It appears in man only when there is a sudden drop in environmental pressure. What is important for this discussion is the fact that the body must have been under high pressure for a long time because it requires a certain amount of time before gases under high pressure can be distributed in the tissues. Of special importance here is nitrogen, small amounts of which are dissolved in the blood and then transported via the circulatory system to the tissues; the tissues adapt slowly to the N_2 pressure. A sudden decompression causes the formation of gas bubbles, some in the tissue itself, some in the blood. The effect now depends on where the gas bubbles are located; among other effects, they can block blood flow. In addition, cells containing fat may rupture.

Nitrogen bubbles which have not been reabsorbed after several hours are especially dangerous. The consequences of this reduction in blood flow naturally depend on the region affected. Air bubbles in the respiratory center produce respiratory paralysis. If other portions of the brain or spinal cord are affected, then motor paralysis can also occur. Air bubbles in the skin ("diver's fleas") are uncomfortable but not lethal. If the bones are affected, then such delayed problems as fractures and bone marrow necrosis can result.

The important thing is that diving accidents caused by caisson disease can be therapeutically treated immediately, before lasting tissue damage occurs. One form of immediate therapy which can be applied is the high-pressure chamber, if one is nearby. In an emergency, it may be necessary to dive again to great depths. The effect of the excess pressure is such that the gas bubbles go into solution once more, after which a slow decompression can be performed. The required time for ascent can be read from tables.

Another danger of deep diving is acute and chronic oxygen poisoning. The higher the P_{O_2} affecting the body, the faster the appearance of acute O_2 poisoning. At the water surface, P_{O_2} is normally about 150 mm Hg when breathing air or compressed air from a diving apparatus; it is 300 mm Hg at a depth of 10 m and 450 mm Hg at 20 m. The upper limit of tolerance is an oxygen pressure of 1,300 mm Hg or 1.7 ata, where duration is also important. The time limit is determined according to the following formula:

$$t = 1{,}000/P^3, \qquad (69)$$

where t is time in minutes and P is partial pressure of oxygen in ata.

At a depth of 74 m, the maximal duration is 3 hours if compressed air is breathed.

If the partial pressure of oxygen surpasses the threshold value of about 1,300 mm Hg, then lung edema will result.

Acute and chronic oxygen poisoning are different. In the case of chronic poisoning, there are primarily pathologic alterations in the respiratory tract (Lorrain-Smith effect), in contrast to which symptoms of the CNS, e. g., convulsions (Paul-Bert effect), are the main effects of acute poisoning. The decisive factor for both forms of poisoning is the product of partial pressure and effective time in the air being breathed.

A further danger is that of depth intoxication ("rhapsody of the deep"), which is also associated with high nitrogen pressure and the dissolving of N_2 into the CNS tissues, causing an effect similar to that of excessive alcohol consumption. The water depth at which this intoxication appears can be increased by replacing nitrogen with helium, which is less soluble.

Diving has unseen dangers which can be avoided only by proper training. It is imperative that divers be advised not to dive to great depths without this training.

4.8.4 Body Temperature in the Water

The heat capacity of water (1 cal $\cdot$ cm^{-3} $\cdot$ degree C^{-1}) compared to that of air (0.31 $\cdot$ 10^{-3} cal $\cdot$ cm $^{-3}$ $\cdot$ degree C^{-1}) is about 3,200 times greater. Thus, one liter water at the same temperature change can take up about 3,200 times as much heat as air. The heat conductivity of water is also 25 times greater than that of air.

The physical characteristics of water are such that the conditions of heat transfer are also basically different from those of air. The survival times of sailors and flyers who jumped into the water during World War II are shown in the diagram of Figure 139. Time in the water is compared to the temperature of the water. It can be seen that all those who survived are to the right of a curve called the tolerance curve. Relatively long survival times can be observed in water temperatures greater than 20 C; there is an acute drop in survival duration in water under 20 C. Remaining in water of 10–15 C for 1–2 hours soon produces death. Water temperatures under 20 C produce hypothermia, i. e., heat production can no longer compensate for heat loss.

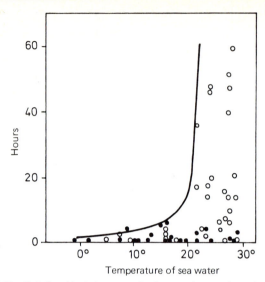

Fig 139   Relationship between water temperature and survival time of flyers or naval personnel during World War II (according to Molnar).

Studies have shown that a naked man in an air temperature of 1 C still has a normal core temperature after 4 hours, whereas the core temperature will drop to 25 C during a 1-hour exposure to 1 C water. During the 1st hour in water of 15 C, core temperature drops 2–3 C, even though the energy metabolism can rise to 5 times the normal level.

Estimates indicate that at 20 C the heat loss in water is about 3 times greater than that in air. This is basically attributed to the greater temperature gradient between body core and environment because skin temperature (with its smaller marginal layer thickness) adapts to that of the water. This threefold increase in heat loss occurs because the temperature gradient between body core and environment increases. Heat loss at an air temperature of 20 C is already 1.5 times greater than that in a thermoindifferent environment. Thus, there is a fivefold greater heat loss in water of 20 C compared to heat loss under basal conditions. This means that 20 C water must represent the critical level of cooling because the metabolism for a prolonged time cannot be increased to more than 6 kcal/minute. Because the temperature gradient between body core and environment is the decisive factor, the effect of fat insulation becomes clear. Heat transfer through the body shell per unit of time corresponds to the product of the temperature difference and the heat transfer coefficient. A smaller heat transfer coefficient can lower the limit of cold tolerance. That is the reason that

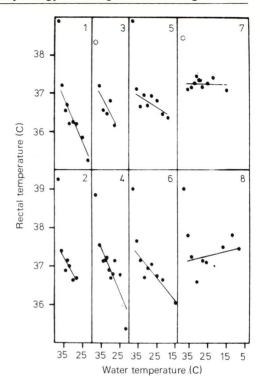

Fig 140   Drop in rectal temperature as function of water temperature with persons of varying degrees of adiposity. Numbers in boxes represent number of subjects; upper boxes correspond to fat subjects and lower boxes to thin ones. From Thauer, R.: Überleben auf See. Kiel, 1965.

fat people can tolerate low water temperatures better than thin people (Fig 140). In the case of higher environmental temperatures which are above that of the body core, the insulation value of the body shell of thin and fat persons is practically the same because alteration in blood flow and not the fat layer now determines the insulation value.

Table 16   Influence of Work on Rectal Temperature and Heat Production of Unclothed Subjects in Cold Water (from Keatinge, K. R.: Quart. J. Exp. Physiol. 46: 69–81, 1961)

Water Temperature Degrees C	Duration min	Number of Subjects	Reduction of Rectal Temperature Degrees C		Heat Production kcal/min	
			Rest	Work	Rest	Work
15	20	12	0.33	0.80	2.84	6.63
5	20	5	1.23	1.81	4.55	6.90

4.8.5 Effect of Swimming on Body Core Temperature in Cold Water

The question can now be raised as to whether an active voluntary elevation in metabolism, such as produced in swimming, can hinder the reduction in body temperature in cold water. Detailed studies have shown that this is not the case. On the contrary, swimming movements appear to accelerate cooling in the same way as uncontrolled shivering. Table 16 shows results from studies in which subjects swam for 20 minutes in water of either 15 or 5 C. Although swimming metabolism in 15 C water was more than double the value obtained under resting conditions, rectal temperature fell faster than it did at rest. It was immaterial whether the swimmer was thin or fat, as there was always a drop in rectal temperature; it was incorrectly believed that this would only occur with thin persons and that there would be a temperature rise in fat persons. This drop can be explained by the fact that peripheral blood flow increases due to exercise, resulting in a marked reduction in insulation of the body shell in fat people as well. The increased heat loss caused by this situation obviously cannot be compensated by the increased heat production of swimming.

In case of an accident in cold water, a person should not throw off clothing because it provides protection in the water by hindering the heat loss due to convection. Wet clothing is also useful for its buoyancy. The region of the neck should be well insulated to delay cooling.

When those who have been chilled are rescued, they should be warmed as fast as possible. Warm baths in hot water are especially good because they produce a rapid warming of body temperature.

4.8.6 Mechanical Work, Energy Metabolism and Efficiency in Swimming

The mechanical work of swimming can only be estimated because it involves overcoming the frictional resistance of water. Frictional resistance itself is a complex variable, which is produced by the sum of skin friction and resistance caused by turbulence and by making waves.

Frictional resistance can be approximated if the force (f) necessary to pull a body through the water at a constant velocity is measured. Resistance increases relative to the square of the velocity (v):

$$f = kv^2. \tag{70}$$

The pulling work is the force (f) applied over the distance (d):

$$w = fd = rd, \tag{71}$$

where w is the work and r is the frictional resistance. If the first equation is inserted into the second, then the propulsive work of swimming is:

$$w = kv^2d. \tag{72}$$

Experimentally determined values for the constants (k) lie between 2.7 and 3.7 kp · sec^2 · m^{-2} for passive pulling on the back or face down. The values of the constants are determined primarily by the size of skin surface involved. When using these formulae, remember that swimmers do not generally swim at a constant speed, but continually accelerate and decelerate. The amount of acceleration and deceleration basically depends on the type of swimming. Another possibility for obtaining information about the propulsive force of swimmers is to have them swim while tethered in a harness and against a cord. If the swimmer is directly connected to a dynamometer whose dial is essentially free of inertia, then the propulsive force (e. g., during the arm pull in the breaststroke) and the negative forces of propulsion (while returning the arm to its initial position) can be measured. With a computer, the positive and negative forces can be separately integrated and the integral of the propulsive force can be quantitatively calculated.

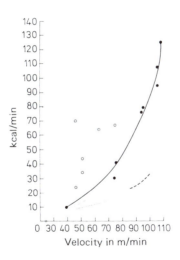

○ Karpovich and Millman (1944) untrained swimmers
● Karpovich and Millman (1944) trained swimmers
···· Pugh et al. (1960) 2 channel swimmers
- - - Unpublished data, Univ Mich., 1965
  (mean of 2 olympic swimmers)

Fig 141 Comparison of data on caloric cost of swimming front crawl stroke at different velocities. From Faulkner, G. A.: Physiology of Swimming and Diving. In: H. Falls (ed.): Exercise Physiology (New York and London: Academic Press, 1968).

The energy expenditure of swimming is understandably difficult to measure without impeding the swimmer. As is the case with all kinds of sport which require a great deal of coordination, the energy expended at the same velocity primarily depends on the degree of practice.

Similar results can be seen in Figure 141, in which the energy requirement is displayed as a function of swimming speed. The energy expended increases disproportionately to the speed. In addition, experienced swimmers have a lower expenditure than untrained swimmers at the same velocity. Estimates of efficiency reveal that the efficiency of swimming can vary from 0.5% for inexperienced swimmers to 8% for well-trained swimmers.

5 Physical Performance Capacity

5.1 Fundamentals

Physical performance capacity is the ability of man to carry out physical actions with his muscles or to cope with heavy forces. The types of action which are performed are so diverse that there can be no single measure for performance capacity. Each individual has a unique capacity for a specific task. This performance capacity can be quantified and compared for the same tasks and the same environmental conditions. If two runners under similar conditions run 100 m in 11 and 12 seconds, respectively, then the performance capacity of the first is greater. Individual performance capacity can also be quantified for a given task in that one person becomes tired and must stop yet another can continue the same task for an unlimited time. Individual physical

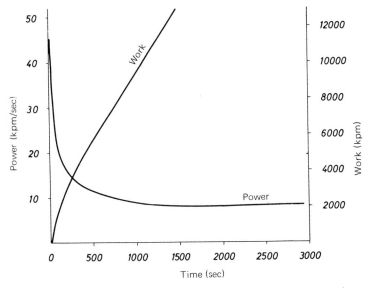

Fig 142   Relationship between power and work. Power curve corresponds to average power of an untrained male subject. Graph shows clearly that maximal total work is theoretically infinite when power required is below threshold for prolonged work.

performance capacity can be properly compared only for the same tasks under the same conditions.

The physical term power describes the amount of work done per unit of time. Figure 142 shows the relationship between power and work. The abscissa indicates time and power in kpm/second is on the left ordinate. The power curve represents values of maximal power which can be maintained within a given time period by a person of average fitness. The curve is a hyperbola approaching both the y axis and the x axis asymptotically. The maximal power that this subject can produce is difficult to determine due to the asymptotic approximation of the y axis. It is 40–50 kpm/second and would be a theoretical value only because the time period is approaching zero. During a time period approaching zero, it is not possible to measure power. The right side of the power curve, in the range where the curve is parallel to the abscissa, indicates that the maximal power is independent of time. This range is the endurance performance capacity. The limit at which a performance can be maintained for an unlimited time period is the threshold for prolonged work or the endurance threshold.

Physical work is the product of power and working time. The curve resulting from this product is labelled work and corresponds to the values on the right ordinate. If high levels of power are generated within a short time, the total work done will be less than that if lower power loads are maintained over a longer period. Working capacity is especially high for power loads which are less than the threshold for prolonged work. This curve is useful when considering performance capacity and power threshold and their practical importance and problems. The maximal working capacity, restricted in this study to dynamic work on the bicycle ergometer, is dependent on work time. From a theoretical viewpoint, it cannot give any information about maximal performance capacity because time must asymptotically approach zero to obtain this value. The threshold for prolonged work is of primary importance for the actual amount of physical work performed.

If this knowledge is applied to the dynamic work generally done in industry, it is apparent that the amount of piece-work done per shift is always greater when the work required does not surpass the endurance capacity of the worker. For this situation, many evaluations should be made, including the total amount of work required per shift, the endurance capacity of the laborers and the most favorable arrangement of work periods and pauses so that the threshold for prolonged work will not be surpassed.

Scientific analysis of the working place and the capacity of the workers is important when there are high physical work loads, especially for piece-work on the assembly line where one is paid for the actual amount of work done. For this reason, workers attempt to do the greatest possible amount of work, even at the risk of using reserves; this may have detrimental effects on their health. For workers paid by the hour, the threshold for prolonged work is seldom exceeded today.

These relationships are different during record athletic performances. The world record for the 100 m run is about 9.9 seconds. The performance required for this would be on the extreme left-hand side of the power hyperbola. At the other extreme is the marathon run (42.1 km), with a record time of less than 2 hours 10 minutes; this would lie on the side parallel to the abscissa. The measurement and evaluation of working capacity are generally limited to physical performances. Performances requiring a great deal of coordination are difficult to determine with physiologic methods. Even with the same working capacity, maximal performance will not necessarily be the same for each individual because psychologic factors are also involved.

Whether the maximal performance carried out during a given time period completely taxes an individual's maximal working capacity depends on motivation. For high-level performances by well-trained athletes, the will to perform will be optimal. Even for assembly line workers, a high motivation can be assumed in men who have no aversion to taking a working capacity test when their salary level depends on it. In addition to the physiologic working capacity, the factors associated with work psychology are very important in determining the actual amount of work done.

Even with maximal motivation, the working capacity of an individual cannot be completely taxed. There are psychiatric cases in which persons reach such a high level of motor stimulation that several attendants have difficulty restraining them. These people are able to increase their performance to levels which are way above their normal working capacity. Persons under life-threatening or highly emotional situations can markedly increase their performance capacity. The well-known classical marathon runner was supposedly so exhausted that he collapsed and died after he had transmitted the message of the victory over the Persians. It appears that under special conditions, reserves are released which are not normally available to one's will to perform.

Therefore, human performance is determined by numerous factors. Two main factors are most important for the actual amount of work performed, i. e., the capacity to carry out a performance and the willingness to apply this capacity.

This discussion will include only the most important findings of research in work psychology and only in so far as they are necessary for understanding the relationships. In the analysis of Figure 143, we start with a fictitious, nondefinable highest level of physical working capacity. If an individual performs at 40% of this fictitious maximal working capacity, he is in the range of automated performances; he is capable of executing the required activity with practically no voluntary exer-

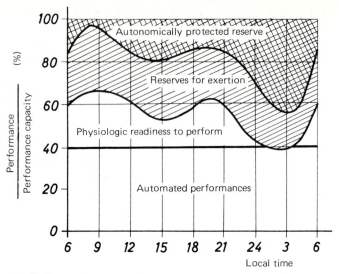

Fig 143   Daily periodicity of readiness to perform (according to Graf).

tion. This range is defined on top by the line parallel to the abscissa. The drive to work that a person is normally ready to apply is called the physiologic readiness to perform. It is defined by a variable curved line, averaging about 50% of the fictitious maximal working capacity. The physiologic readiness to perform changes with the time of day. The pattern of this curve corresponds to the general experience that work feels much easier in the morning hours then during the early afternoon or late at night. This applies regardless of eating times or sleeping patterns. The reason for this phenomenon is the circadian rhythm, which is controlled by the autonomic nervous system. This circadian rhythm can also be seen in other autonomic variables, e. g., the 24-hour pattern of core temperature or heart rate. The cause of these patterns is not known but circadian rhythms supposedly occur due to an internal clock, which is synchronized with such external factors as daylight. They are closely dependent on local time. When the local time is shifted by a sea voyage or long flight, there is a phase displacement. Complete adaptation to the new local time occurs with time; the adaptation will usually be corrected about 2 hours each day. Airline flight personnel are especially bothered and some report such unpleasant reactions as sleep disturbances or reduced working capacity. When athletes are not flown to the site of competition in sufficient time for adaptation, serious disturbances in performance result. The fact that the curve is dependent on local time is especially important for workers on the night shift, in whom the circadian rhythm remains the same, despite the reversal in living habits.

The hatched area in Figure 143 shows the reserves for exertion which can be voluntarily mobilized and which are influenced by circadian rhythms. During a high-level performance (sports), the usual reserves for exertion can be mobilized up to 80% of the fictitious maximal working capacity. The rest of the performance reserves are autonomically protected. They are not accessible to one's will and can be released only by strong emotions (danger to life) via the secretion of epinephrine. These last performance reserves can be made available by using drugs similar to epinephrine which by-pass the normal pathway. There is naturally a great danger for the body with these procedures (doping). These effects must not be confused with a true improvement in physical performance capacity, which pharmacologically is not possible.

Even though the readiness to perform is a complex psychophysiologic variable and is subject to variation depending on the time of day, physical performance capacity is relatively constant, at least over brief time periods. The major problem is to measure it with sufficient accuracy; a completely satisfactory procedure is not yet available. The simplest method is to give a subject a defined work load on a bicycle ergometer, which is assumed to be greater than the threshold for prolonged work, and then to measure how long he can maintain this work load. However, even with this simple test, there are difficulties. Physical performance capacity measured in this way is dependent on motivation and requires maximal cooperation by the individual. In addition, the tests must be given at the same time of the day to be certain that the readiness to perform remains somewhat constant. These problems make such performance tests unreliable, at least in the area of industrial work, because the assistance of the person being studied is perhaps uncertain. This is entirely separate from the fact that it is dangerous to stress older or latently sick workers up to the limits of their performance capacity.

Testing procedures now use reactions of the circulatory and respiratory systems to a work load as criteria for physical performance capacity. The classical procedure, which will be discussed in the next section, is always dependent on determining the weakest link in the cybernetic system. For an activity which is uniform and undifferentiated, such as bicycle ergometer exercise, it is simple to determine the weakest link, which is the oxygen transport system within the organism. However, the more complicated the work task, the greater the emphasis will be on other factors which limit physical performance capacity. They are not so easily determined with physiologic testing procedures and are understandably modified by environmental conditions and the CNS.

Therefore, all testing procedures must be carefully and critically analyzed.

5.2 Maximal Performance Capacity

Blood flow can no longer transport the necessary oxygen for aerobic metabolism to the muscle if the local blood flow cannot be further increased or if the pumping capacity of the heart has reached its upper limits. In the latter case, maximal cardiac output is the limiting factor and can be measured only with difficulty since a large amount of equipment is necessary. For this reason, a factor closely related to it, the maximal oxygen intake, is determined. This gives an estimation of the range of adaptation of the circulatory system. The enzyme content in the mitochondria is another important limiting factor.

5.2.1 Aerobic Power (Maximal O_2 Intake)

When a high work load performance is desired, during which many muscle groups participate, the oxygen requirement will be greater than the maximal O_2 intake. When a high work load causes fatigue within several minutes, 98% of the maximal O_2 intake is reached after 3–4 minutes (Fig 41). To understand why maximal O_2 intake and maximal working capacity are related and why maximal O_2 intake is increased with training and is reflected in the increase in working capacity, the parameters limiting or influencing maximal O_2 intake must be considered. This discussion will be limited to healthy man. The first important variable is maximal cardiac output, which is dependent on maximal stroke volume and maximal heart rate. Because all oxygen which passes from the lungs into the blood is transported further in the blood, oxygen intake must level off when the AVD for oxygen at this maximal cardiac output is fully used. Thus, the other factor which determines maximal O_2 intake is the AVD for oxygen in the blood, also called peripheral extraction. This extraction is dependent on several factors, including capillarization of the muscle and capillary blood flow velocity. Good capillarization produces a larger diffusion surface and better oxygen extraction. The blood returns to the lungs with a lower oxygen content and can be replenished to a greater extent. Total peripheral extraction is primarily determined by the organ system through which the blood flows. For this, enzyme content and capillarization are decisive. Blood will be much better utilized in well-capillarized muscle than in well-capillarized skin because it has a nutritive function in the muscles and a heat-regulating function in the skin. At the same maximal cardiac output, if there is a greater blood flow to the skin, maximal O_2 intake will be lower. An increased capillarization of muscle caused by training, combined with an increase in aerobic enzymes, would increase maximal O_2 intake.

Maximal O_2 intake is also affected by the hemoglobin concentration. Hemoglobin concentration can increase (adaptation to altitude) or the blood volume can increase with a slight decrease in hemoglobin concentration (endurance training); in both instances, there is an increase in the total amount of hemoglobin. A well-trained athlete has a reduced concentration which is about 3% less than the normal value, but has a 15% increase in blood volume. Blood volume is more effective because there would be a rise in blood viscosity when there is too great a hemoglobin concentration due to the increase in erythrocyte count. As a result, the amount of work done by the heart would be greater and the maximal values of cardiac output and O_2 intake would be reduced.

The average total extraction from the blood during maximal performance is estimated at about 60%. The AVD for O_2 at a hemoglobin concentration of 16 gm/100 ml is about 120 ml O_2 per liter blood; this means that effective perfusion per minute through the lungs can produce a maximal O_2 intake of 25×120 ml = 3 l/minute. If the cardiac output could be increased by training to a maximum of 35 l/minute, this alone could cause an increase in maximal oxygen intake of about 1.2 l/minute to a total of 4.2 l/minute. If there is an increase through training in peripheral O_2 extraction of 5% to 65%, corresponding to an increase of 10 ml O_2 per liter blood, then the maximal O_2 intake could again be increased by about 350 ml/minute.

For sick people, maximal O_2 intake can be limited by still other parameters which, at least in normal young individuals, are always greater than required, such as lung ventilation and capacity for diffusion of oxygen across the alveolar wall.

5.2.2 Determination of Maximal O_2 Intake

The determination of maximal O_2 intake assumes a high level of physiologic stress, to which only healthy persons should be subjected. A previous medical examination is required. Primary contraindications include a myocardial infarction within the last 12 months, coronary insufficiency, hypertension or any other type of resting cardiac insufficiency. Any investigation should be terminated if frequent cardiac rhythm disturbances appear.

The measurement technique depends on the method by which O_2 intake is determined. Scandinavian investigators measure maximal O_2 intake during work on a treadmill; during this test with adults, there is a speed of 10–14 km/hr and a grade of 10%. An untrained person will be exhausted after 4–6 minutes. At least 2 minutes after the onset of exercise, expired air is collected in two Douglas bags during each minute (Suppl, p 309). The difference in O_2 intake between the two samples should be small because after 2 minutes of exercise, 90–98% of the maximal O_2 intake will be reached; after 3 minutes, this will be 95–98%. Maximal O_2 intake, or the vita-maxima, as it is primarily measured using the closed system (Suppl, p 313) in the clinic and in sports medicine, is somewhat different. With untrained subjects, it is common to test on the bicycle ergometer at 2–6 kpm/second (depending on the age, sex and constitution of the subject) and then to increase the working intensity every 3 minutes by 2–3 kpm/second. Because oxygen intake can be continually measured in this instance, the increase in O_2 intake can be determined with each increase in work, during which the subject reaches a new steady-state after 2–3 minutes. When the work must be interrupted, the maximal O_2 intake is obtained. This standard test is also possible using the continuous O_2 determination method (p 313). To compare the methods, it must be remembered that in the closed system, the integral is measured during the last minute, whereas the actual value at exhaustion is determined during breath-by-breath analysis. A corresponding transformation is possible.

The determination of maximal O_2 intake in athletes produces additional problems. The first is associated with the varying levels of conditioning of the muscle groups that are used during the special sport discipline. If maximal O_2 intake is determined on the bicycle ergometer, then those muscle groups that are important for the performance of a cyclist will be stressed. If a long-distance runner is tested in the same way, he will show lower values because he is stressing muscle groups that are only partially trained. In this case, maximal O_2 intake should be measured on the treadmill (Fig 43). The optimal procedure would be to develop a special type of ergometer for each type of sport.

Another problem occurring during bicycle ergometer exercise with well-trained athletes is that possibly local muscular endurance and not the cardiac output represents the limiting factor. According to previous investigations on untrained persons, local muscular endurance should be limiting when less than one-seventh of the musculature participates; the circulatory system should be a limiting factor when more than this proportion performs work.

5.2.2.1 Criteria for Adequate Exertion by Subjects

For the correct determination of maximal O_2 intake, it is assumed that the subject is adequately stressed. Numerous criteria can be used to determine whether the subject was actually exhausted or whether he gave up the exercise prematurely due to a lack of motivation. One reliable criterion is the ventilatory equivalent, which refers to the relationship between ventilation and O_2 intake ($\dot{V}_E/\dot{V}O_2$). At rest, the ventilatory equivalent has a value of about 20–25. As a result of the disproportionate respiratory drive at work loads greater than the threshold for prolonged work, it increases and may reach a value of 40–50 at exhaustion because the rise is especially steep shortly before exhaustion. The end-tidal CO_2 pressure can reach a value of about 30 mm Hg at this time.

The maximal heart rate reached at exhaustion depends primarily on the age of the subject and is only a relative measure. It is recommended that not just one criterion be used, but that a combination of all of these should be used. This requires a certain amount of experience, however.

5.2.3 Calculated Maximal Aerobic Power

The dangers for older and sick persons when testing the circulatory system and the difficulties occurring with unmotivated subjects have led to other types of procedures, in which the maximal O_2 intake is not measured but the values of heart rate and O_2 intake during submaximal exercise are extrapolated to the maximal O_2 intake. The advantage is that subjects no longer must be overly stressed. However, significant errors can appear. The nomogram in Figure 144 is constructed from measurements of O_2 intake, heart rate and maximal O_2 intake of a large sample of subjects, primarily young men. The relationship between heart rate and O_2 intake during submaximal work was so closely related to maximal O_2 intake that this relationship can be assumed with sufficient accuracy for other subjects. This assumes that the sample used in this nomogram is, on the average, the same as that sample which will be tested in the future; this always represents a significant factor of uncertainty. Physiologically, the nomogram is based on the fact that heart rate is controlled by peripheral requirements (p 133) and that cardiac output automatically adapts to venous return. At the maximal attainable heart rate (assumed to be 195 beats per minute for all subjects), cardiac output and maximal O_2 intake are both assumed to be at their maximum, provided blood volume has increased harmonically. Thus, heart rate and oxygen intake must be relatively closely related in young men. However, in older men, just the fact that the maximal attainable heart rate decreases due to age means that significant deviations will occur between the O_2 intake determined from the nomogram and the actual measured maximal O_2 intake. The reduction in maximal O_2 intake due to the decrease in maximal heart rate can be determined from Table 17.

Another method has been developed by Scandinavian researchers. Maximal O_2 intake can be determined according to the following relationship:

$$\text{max } \dot{V}O_2 = 1.29 \sqrt{\frac{L}{H\text{-}60} \cdot e^{-0.00884\,A}}, \tag{73}$$

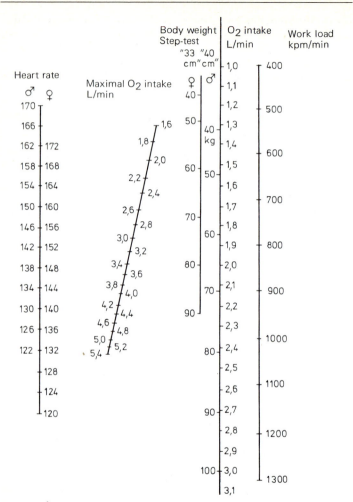

Fig 144   Åstrand-Ryhming nomogram for determining maximal O_2 intake.

where L is the submaximal work load in kpm/minute, H is the heart rate after 5–6 minutes at the work load L and A is the age in years.
The maximal O_2 intake of women is normally less than that of men due to their smaller heart size. They show similar changes with age and the same per cent increase due to training.
It is more common today to express maximal O_2 intake in ml/kg body weight per minute because it will significantly reduce variations. A

Table 17   Correction Factors to be Used with Nomogram of Figure 144 for Different Age Levels in Both Sexes (according to W. Hollmann)

Age (yrs)	Correction Factor
25	1.00
35	0.87
45	0.78
55	0.71
65	0.65

Table 18   Average Values for Maximal O_2 Intake for Trained and Untrained Persons (M male, F female).

	Max. O_2 l/min	Max. O_2/ ml/kg min
Trained (M)	4.8	67
Untrained (M)	3.2	44
Trained (F)	3.3	55
Untrained (F)	2.3	38

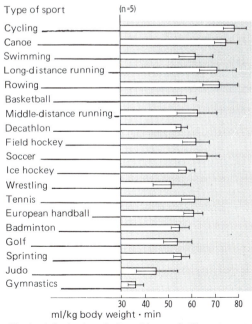

Fig 145   Maximal O_2 intake of elite athletes of different sports (according to data of Hollmann and Heck).

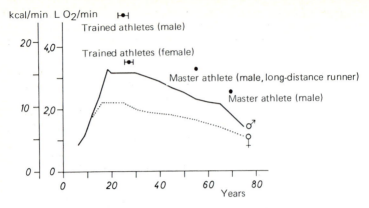

Fig 146 Maximal O_2 intake of untrained persons as function of age and sex. For comparison, data from endurance athletes are also included (according to Hollmann).

small slender runner with a maximal O_2 intake of 4 l/minute has a large working capacity in comparison with an athlete weighing 90 kg with the same measured values. Normal values for young trained and untrained subjects can be seen in Table 18.

As can be seen in Figure 145, where the average of 5 top athletes in each type of sport is recorded, the definition of "trained" is nonuniform. Maximal O_2 intake is essentially greater in the typical endurance sports than in those requiring a great deal of coordination. Figure 146 shows the relationship of maximal O_2 intake and age for both sexes.

5.3 Endurance Performance Capacity

Maximal performance can never be the decisive factor for performance in those occupations requiring physical work. Much more important is "lifetime efficiency," i. e., the total amount of work performed over the entire life span. This assumes that the body is not continuously stressed at levels greater than the threshold for prolonged work because injuries will then occur. These have been labeled under the collective term of "wear-and-tear diseases." How some of these individual injuries are causally related to chronic overload is still unclear, especially when results are compared with findings of modern gerontologic research. The causes are probably related to the processes of molecular biology during cell reproduction. A tissue which has a high metabolic rate or which has been mechanically overloaded must be regenerated many times. It is assumed today that genetic information is lost during this regeneration or that the matrix is worn out. In addition, the residuals of metabolic end products remain in the cells and cause premature cellular aging.

Inadequate stress of the organism also produces other diseases, which have been labeled "civilization diseases". The primary cause of these diseases is the fact that the organism has lost its ability to adapt to sudden stresses because of this inadequate stimulation. Circulatory diseases and myocardial infarction are examples of diseases caused by this chronic physical inactivity, even though they are not exclusively associated with physical inactivity.

Researchers have long sought an objective measure for the threshold for prolonged work (also called the endurance limit). At first, it was suggested that this should be the work load which can be performed at 50% of the maximal O_2 intake. However, 50% of maximal O_2 intake could be a gross and occasionally false measure for the threshold of prolonged work. This is so for the following reasons. Young men can maintain work loads which correspond to these requirements for relatively long periods of time. In older men, endurance capacity approaches maximal performance capacity because the maximal performance capacity drops off more than the endurance capacity. In addition, depending on the type of occupational requirements, the performance limits may be at a level of oxygen intake which is perhaps only 20% of the maximal O_2 intake. In this case, it is not the work capacity of the heart but the local blood flow to the muscle which limits the capacity to work.

Another variable is needed to give information about the threshold for prolonged work. The heart rate is an almost ideal indicator because it is an indirect measure of muscle blood flow. Because the physiologic bases have already been discussed (p 123), this section will be limited to the practical consequences, the usefulness and the predictive value of this method.

Changes in heart rate during work below the threshold for prolonged work are shown in the upper half of Figure 147. Heart rate rises with the onset of exercise and attains a steady state after a short period of time. After the exercise is over, it falls relatively quickly back to its resting value. If the work load is greater than the threshold for prolonged work, then there is an initial rapid rise in heart rate, no steady state is reached and heart rate increases continually during additional exercise. After exercise, the heart rate shows a delayed return to its initial value. The threshold for prolonged work can be determined from two characteristics of heart rate, the changes during exercise and the changes after exercise. The heart rate pattern during exercise under laboratory conditions is a more appropriate way of determining the level of the threshold for prolonged work. Measure-

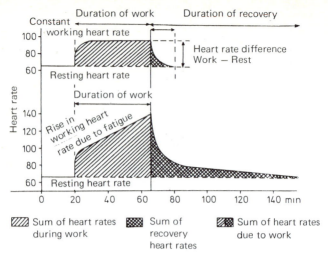

Fig 147   Heart rate during and after work of different intensities (according to E. A. Müller).

ment of the heart rate pattern after exercise is especially useful in determining whether and by how much the endurance performance capacities have been overloaded.

The threshold for prolonged work can be estimated initially with the performance heart-rate index (p 246) or the work load on a bicycle ergometer can be varied until the heart rate of the subject is 120–130 beats per minute. Both methods reduce the range in which the threshold for prolonged work is found. In a strong man, the estimated value is 11–13 kpm/second. He can perform a work load of 11 kpm/second for 15 minutes on the bicycle ergometer. The results are statistically calculated so that the regression between time and heart rate from the 4th to the 15th minute is calculated. If the resulting correlation coefficient is under 0.8, then the work load is less than the threshold for prolonged work. The next work load would then be 12 kpm/second. If the correlation coefficient of the corresponding heart rate to time regression is greater than 0.8, then the threshold for prolonged work has been surpassed. This method can be used to narrow the threshold for prolonged work of a subject to within about 0.5 kpm/second for bicycle ergometer exercise. This method is very time-consuming as regards the statistical calculations if no desk-top computer is available. The increasing availability of these calculators makes the present method a useful and simple test for the threshold of prolonged work, which can be described in kpm/second for bicycle ergometer exercise (at 60 rpm). Work load must obviously be related

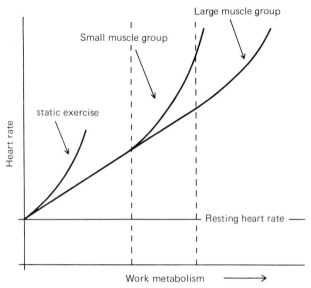

Fig 148   Heart rate with increasing muscle metabolism.

to an apparatus such that net metabolism and work load are closely related. If this is done, the limiting factor is never directly the work load performed but the maximal aerobic metabolism in the musculature. For these reasons, it is not sufficient just to use the term "bicycle ergometer exercise;" the number of pedal revolutions must be standardized because the mechanical efficiency does not change with work load but changes with the number of pedal revolutions. All data are usually standardized to 60 rpm.

The second informative factor for the threshold for prolonged work is the sum of recovery heart rates (p 250), even though this is much more difficult to determine precisely. The rise in heart rate during exercise and the sum of recovery heart rates give an exact measure of the threshold for prolonged work of the muscle if the local blood flow limits the performance. This occurs because of the mechanism for heart rate adjustment. Figure 148 shows the changes in heart rate during work with a small muscle group in comparison to work with a large muscle group. If a small muscle group is working, then at the same absolute work load the maximal blood flow will be attained at a smaller work load. Therefore, the increase in heart rate will also lose its linear relationship to the increased oxygen intake at lower values than occurs with larger muscle groups. Figure 149 shows the average endurance limits of men up to the 7th decade.

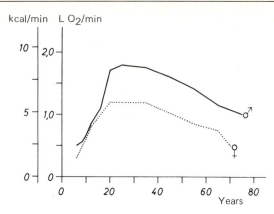

Fig 149   Average thresholds for prolonged work of untrained persons as function of age and sex (according to Hollmann).

5.3.1 Estimation of Performance Capacity from the Pattern of the "Oxygen Pulse"

Oxygen pulse is defined as the amount of oxygen which is taken up per heart beat. To obtain the oxygen pulse value, determine the oxygen intake (ml/minute) and heart rate (beats per minute) over a period of 5 minutes and then divide the first variable by the second. To keep the mathematic relationship simple, it is shown schematically in Figure 150. If the linear relationship between increase in heart rate and oxygen intake is extrapolated to the left, then at an oxygen intake of 0, the y axis would transect a positive value, i. e., a heart rate which is well over 0. This range to the left of the vertical line, which is supposed to represent resting metabolism, is only of theoretical importance. Within this range, the oxygen pulse must have a value of 0 when it crosses the y axis. It must rise with increasing oxygen intake, a fact which is produced simply by the formal relationship. Even though there is a linear function between increased heart rate and increased oxygen intake, the oxygen pulse ratios show a nonlinear relationship to work load. This fact makes it easier to estimate indirectly the performance capacity of a subject.

5.3.2 Estimation of Performance Capacity with the "Performance Heart-Rate Index" of Müller

The performance capacity of untrained persons can also be estimated using the method of the performance heart-rate index (Leistungspuls-index = LPI) of Müller. Compared with other methods, information

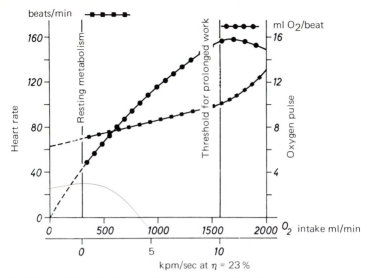

Fig 150   Relationship between heart rate as function of power, oxygen intake as function of power and oxygen pulse.

can be obtained with a small amount of apparatus. As long as the number of pedal revolutions remains constant, the mechanical efficiency is independent of work load on the bicycle ergometer; it remains constant at about 20–22% up to the highest work loads. If the pedal revolutions remain constant during the test, it is not necessary to measure the oxygen intake. The simplest way to determine the LPI is to use Müller's bicycle ergometer. This ergometer has a built-in mechanism which automatically increases the work load each minute by 1 kpm/second.

After a preliminary rest period of 15 minutes, the subject sits on the bicycle ergometer and pedals at 60 rpm against no resistance. The magnetic brake becomes activated after 2 minutes, so that there is an adjustment of 1 kpm/second after the 3rd minute. The test is completed after 12 minutes with a work load of 10 kpm/second. Heart rate is determined during the entire work period. If the average heart rate is determined for each minute, then the graph appears as in Figure 151. The angle of rise in heart rate can be determined either using a straight line or by a simple linear regression calculation using the rules of statistics. The greater the work capacity, the lower the LPI. Mean values and the frequency distribution of LPI for untrained subjects are seen in Figure 152 for both sexes.

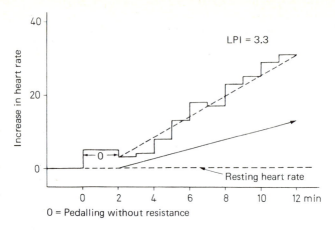

Fig 151   Performance-heart rate index (LPI) of E. A. Müller.

The LPI may be adequate to differentiate the more fit among a group of untrained subjects. However, the LPI is completely inadequate to evaluate the working capacity of men who have improved their physical working capacity through training. Even with a high level of endurance training, there is little change in the LPI although the threshold for prolonged work, as well as the maximal O_2 intake, can be as much as 50% higher than in the untrained state. There are highly trained cyclists whose LPIs are not lower than those of untrained office workers, even though at the same work load which the cyclist is able to maintain for over 1 hour, the untrained men are exhausted after 5 minutes (Table 19).

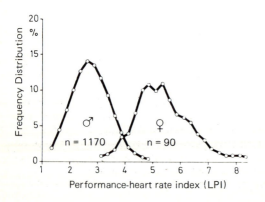

Fig 152
Normal distribution of performance-heart rate index (LPI) of men and women.

Table 19   Comparison of E. A. Müller's LPI among Various Occupational Groups and Well-Trained Cyclists (according to H.-V. Ulmer)

E. A. Müller		"Normal Men"		LPI = 2.8 ± 0.6	
Roth		Police Officers		LPI = 3.0 ± 1.0	
Kaminski		Forest Workers		LPI = 3.1 ± 1.0	
H. Scholz		Foundry Workers		LPI = 3.2 ± 0.8	
Subj	VP	Qualification		n	LPI ± sd
1	UH			2	3.09 0.58
2	KH	1968 Olympics		2	2.40 0.17
3	JK	4-Man Pursuit		3	2.29 0.41
4	KL			8	2.40 0.33
5	AI	Second in Germany, 1968		5	2.71 0.18
6	IR	4-Man Pursuit		2	2.98 0.40
7	MS	German Tandem Champion, 1966–68		2	3.09 0.09
8	ED	4th Place	Year's Best	3	3.25 0.38
9	GS	17th Place	Performance	3	3.30 0.36
10	WT	65th Place		7	3.03 0.17
LPI-Average					2.85 ± 0.37

During the determination of the LPI, incorrect information can be obtained, especially when inexperienced subjects are used for these types of test. Heart rate can be higher during the phase of pedaling against no resistance due to the excitement. As a result, a high performance capacity can be simulated. Whenever possible, it is always recommended that the test be given on at least 3 different days. Nicotine and caffeine taken before the test may also produce false results.

5.4 Muscle Fatigue

Each work load above the threshold for prolonged work causes a reduction in performance capacity. This has been termed muscle fatigue or peripheral fatigue. As has been discussed (pp 44 and 242), the assumption for work which is not limited by muscular factors is that there will be a steady state (continuous equilibrium) between energy supply and energy requirement. As shown in Figure 34, ATP concentration essentially remains constant up to this limit because it is restored via the reserves of creatine phosphate partially during and partially after the contraction. Because the drop in energy-rich phosphates directly controls aerobic and anaerobic metabolism, the level of creatine phosphate decreases, depending on the level of the metabolism. Within this range, the reduction and the replenishment are equilibrated at reduced concentrations. Therefore, the muscle within this range is theoretically able to continue working for an unlimited time.

If the energy requirement can no longer be met aerobically, then the anaerobic energy supply begins and fatigue sets in. The cause of muscle fatigue is the fact that the physical-chemical equilibrium becomes so disturbed that the local energy supply is exhausted after a certain time. Under conditions of reduced work load or at rest, recovery occurs; this is essentially a reinstatement of equilibrium.

5.4.1 Measurement of Muscle Fatigue

The sum of recovery heart rates is a good measure of muscle fatigue. In this regard, recall the change in heart rate described on page 244 during mild (nonfatiguing) and heavy (fatiguing) work loads on the bicycle ergometer (Fig 147). We are now primarily interested in the change in heart rate after exercise. In the case of mild exercise, heart rate reaches its resting value in 2–3 minutes. The exact mathematical analysis of the return in heart rate to resting values shows that it tends to return with a simple negative exponential function. In the case of fatiguing exercise, the function is at least the sum of two exponential functions. Several hours may be required for the complete return. The number of heart beats which are greater than the initial resting value can be determined during the period from the end of exercise to the point where the resting value is once again attained. Figure 153 shows an example of this measurement. The previous exercise was not fatiguing if the sum of recovery heart rates is equal to or less than 100. Values for the sum of recovery heart rates of several thousand can occur after fatiguing exercise. Detailed investigations have shown that the degree of fatigue is proportional to the product of the work performed which is greater than the threshold for prolonged work and the time of exercise.

5.4.2 Muscle Fatigue and Work-Rest Ratios

When the biochemical imbalance is compared with the degree of fatigue, important consequences are found for the rhythm between work and rest phases. Two factors influencing recovery are the removal of metabolic end products and the resynthesis of the degraded substances. The Fick diffusion law applies to the first reaction, i. e., diffusion velocity is proportional to the concentration differences under otherwise identical conditions. The concentration difference between the cell, extracellular space and capillary reaches a maximum immediately after exercise; this maximum becomes less with time. If the blood were to stay in the capillaries, the equilibrium would follow

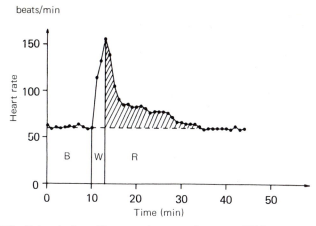

Fig 153 Determination of the sum of recovery heart rate (EPS, striped area). B, period before, used to determine resting heart rate. W, work (in this case, a work load of 24 kpm/second for 3 minutes). R, period of recovery for determination of EPS. Heart rate is dotted line, resting level is dashed line.

exponentially. In the case of flowing blood, the reduction in concentration within the muscle corresponds to an essentially exponential function. Similar regularities also apply to the resynthesis of energy-rich compounds and obey the law of mass action. This means that the speed of reaction, under otherwise similar conditions, depends on the concentration product of the reaction partners. Synthesis immediately after exercise is faster because the concentration product is at its maximum. The reaction velocity drops exponentially. Both variables change in such a way that the recovery value of a pause also decreases exponentially. The onset of the pause is thus more effective for recovery than the end of a prolonged pause.

The results of experiments on the distribution of work and rest periods on the bicycle ergometer are shown in Figure 154. This figure shows the changes in heart rate when the same work load of 20 kpm/second performed on the bicycle ergometer. The upper curves are produced when there are two 5-minute work periods, interrupted by a pause of 7.5 minutes. Here, the heart rate reaches high values. As a measure of the degree of fatigue, the sum of recovery heart rates during the total work of 12,000 kpm shows high values. It is more advantageous if the same amount of exercise is done in 2 minutes with 3-minute pauses interspersed. The total amount of work which can be performed until exhaustion is more than doubled. However, the sum of recovery heart rates is not much greater. Even more advantageous conditions are shown in the lower part of the figure, during which 30 seconds of

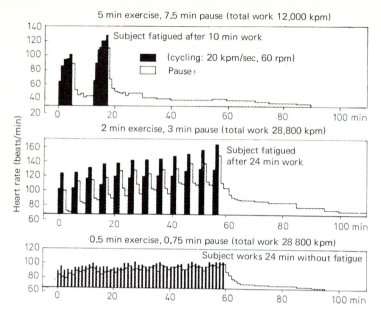

Fig 154   Effect of varying distributions of work and rest intervals on fatigue, as measured by heart rate (according to E. A. Müller and K. Karrasch).

exercise is alternated with 45 seconds' rest. For the same total amount of work, the subject is not fatigued and the sum of recovery heart rates is extremely low. The theoretical conclusion is that shorter intermittent pauses are essentially more effective for recovery than pauses of prolonged duration.

5.4.3 Recovery by Changing Muscle Groups during Dynamic Work

If a high work load is performed with a muscle group in a manner producing local fatigue, then this muscle group requires an increased blood flow during the recovery phase. It is interesting to know whether the perfusion of the previously fatigued muscle group will be reduced if another muscle group begins to work during the recovery phase. In addition, it would be interesting to know whether the perfusion of the working muscle group is reduced due to the increased perfusion of the previously fatigued muscle group. This is illustrated in Figure 155. A subject works on an arm-crank ergometer at 7.5 kpm/second. This is a work load which he can perform because after 10 minutes of exercise, a

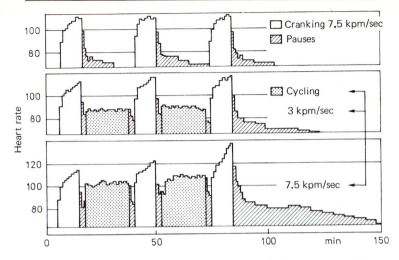

Fig 155   Recovery from cranking exercise by cycling during the pauses. From E. A. Müller: Die physische Ermüdung. In: Handbuch der gesamten Arbeitsmedizin, vol 1 (Munich: Urban and Schwarzenberg, 1961).

25-minute pause is sufficient for complete recovery, as seen in the heart rate which has returned to its resting value. Instead of resting during the recovery phases, the subject then performs 20 minutes of leg exercise on a bicycle ergometer at work loads of 3 kpm/second and 7.5 kpm/second. The figure shows the results. The balance between fatigue and recovery, which was obtained during arm-cranking exercise by a 25-minute pause, is now disturbed during the bicycle exercise, depending on the work load, i. e., the higher the work load, the more the recovery is disturbed. The sum of the recovery heart rates after the 3rd period, which was 137 beats in the upper graph, increased in the intermittent pauses to 224 with the bicycle exercise of 3 kpm/second. During bicycle exercise of 7.5 kpm/second it increased to 846 beats. The heart rate during bicycle exercise was thus added to that which remained from the cranking exercise. Cycling diminished the recovery from cranking, which can be seen from the absence of a drop or a small reduction in heart rate during the period of cycling. This is also seen in the increasing number of heart beats from cranking period to cranking period, as well as by the increased sums of recovery heart rate after the last cranking period. The increased perfusion of a large exercising muscle group thus reduces the perfusion of other organs. Included in this are the other muscles which are at the time in a recovery phase and have an elevated resting blood flow. The body considers the task of recovery to be less important than the task of exercise occurring at that moment.

5.5 Work Capacity and Fatigue during Static Work

Static exercise is an activity during which a weight is held by muscle. Even though increased amounts of energy are necessary, no external work (force × distance) is performed. Static work can be compared to the work done by a helicopter when it hovers in the air. In this case energy is being utilized but the external work is zero. Using the isolated gastrocnemius preparation, dog muscle can develop a maximal force of 6 kp/cm^2 muscle cross section with maximal stimulation of the nerve. If the metabolic regulation of vascular diameter is bypassed with drugs, such that the smallest arteries are at their maximal width, then the resistance to flow during artificial perfusion can be measured as a function of static strength (Fig 156). The resistance increases

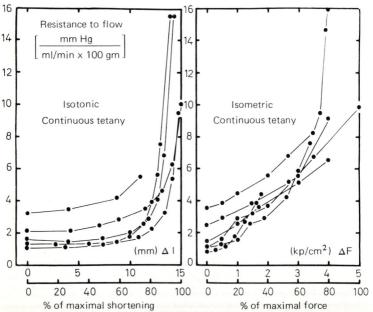

Fig 156 Effect of isotonic and isometric continuous tetany on resistance to flow after pharmacologically-blocked blood flow. From Hirche, J., Raff, W. K., Grün, D.: Pflügers Arch. 314 : 97, 1970.

markedly with static strength during continuous isometric tetany because the capillaries are squeezed due to the increased diameter of the muscle fibers. The greater the initial stress of the muscle, the greater will be this effect. The inhibitory effect is significantly less during rhythmical isometric work. As is the case in all types of muscle contraction, metabolism of the muscle increases markedly with isometric contractions as well. However, as a result of compression, blood flow cannot increase as much as is required by the contracting muscle. Therefore, even at lower levels of energy, there is a disparity between the requirements and the transport of oxygen, compared to the case during dynamic exercise.

If the practical relationships between holding time and static strength are examined, a uniform relationship will be found for each muscle (Fig 157). On the abscissa is static strength in per cent of maximal strength of the muscles being studied; maximal holding time in minutes is on the ordinate. The greater the utilized force in comparison with the maximal force, the greater will be the reduction in holding time. If the strength is less than 15% of the maximal strength, then force is not related to holding time. Tensions which are less than 15% maximal strength can be maintained for unlimited periods. Oxygen supply is the same as the oxygen requirement of the muscle.

Due to the fact that the oxygen supply is no longer identical to oxygen requirement at tensions greater than 15% maximal strength, holding duration is limited. The metabolic end products remain in the muscle and cause an increase in heart rate via the muscle receptors and the circulatory center.

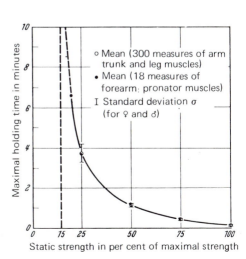

Fig 157
Relationship between static strength and holding time. From Rohmert, H.: Int. Z. angew. Physiol. 18.123, 1960.

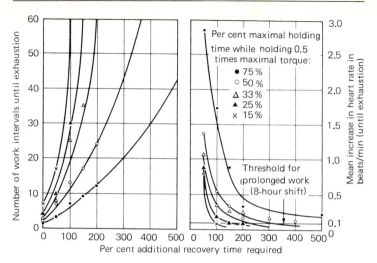

Fig 158 Additional recovery time required in relation to number of work intervals which produce exhaustion (according to Rohmert).

Fatigue increases during constant static strength in a way that is disproportionate to holding time. This relationship can be measured when holding time is compared with the length of the pauses necessary to eliminate residual fatigue. The percentage relationship between the length of the work period and the length of the recovery interval is called the additional recovery time required to compensate for exercise. Figure 158 shows the results of studies in which work periods and pauses were alternated. The task was to hold 50% of a maximal torque. This tension can be maintained for about 1 minute. The different parameters show the values for the number of work phases wich could be maintained until exhaustion (*left*) and the average increase in heart rate (*right*) when the holding time was 75%, 50%, 33%, 25% and 15% of the maximal holding time. When holding time was doubled from 25 to 50% of the maximal holding time, the recovery pauses had to be increased from 150% to 400% of the holding time to continue performing the exercise without fatigue. When the holding time was tripled to 75%, then the length of the recovery pauses had to be increased to a value which was 7 times as great as the holding time. Static holding work is essentially more fatiguing than dynamic work and should be avoided if at all possible. This is evident in the disproportionate heart rate relative to O_2 intake and can always be seen when exercise contains a high proportion of static work.

5.6 Reduction of Work Capacity due to Central Fatigue

In contrast to muscle fatigue, whose cause is found in the reduction of energy-rich compounds or in the appearance of metabolic end products in the muscle itself, there is a phenomenon which is termed central fatigue. It is primarily a decline in the ability to carry out coordinated movements with the same precision as in the nonfatigued state. Central and peripheral fatigue are often linked together. On the other hand, the presence of a pure central fatigue can also be established, as during prolonged observational tasks or during psychic stress. The term central fatigue encompasses a psychophysical complex of symptoms, the cause of which is not yet completely known. Subjectively, it is frequently linked with a feeling of tiredness. The field of work psychology has studied numerous symptoms generally attributable to a decline in central nervous performance. Disturbances in reception have been found, e. g., flicker-fusion frequency is reduced and the upper limit of hearing frequency decreases. Visual perception disturbances during fatigue can be partially related to diminished coordination of the eye muscles (double images). Peripheral coordination disturbances, in turn, amplify peripheral fatigue because the metabolic requirement for the same work load increases in the uncoordinated state. Disturbances in attentiveness, concentration and thinking have also been observed. Reaction time is prolonged, especially when there are several possible alternatives. Motivation and control functions are reduced (fatigue of will).

Closely associated with central fatigue is the reduction in performance capacity due to a stimulation effect which is too small (monotony), as can be found during work on an assembly line, as well as during repetitive supervisory tasks. This "overstress through understress" produces symptoms which are similar to the central fatigue after strenuous muscular exercise.

There is still much uncertainty about the causes. The feeling of tiredness is obviously associated with the function of the reticular formation. When stimuli enter from sensory or sensitive pathways, not only are specific responses initiated in certain areas of the cerebrum, but the reticular formation is simultaneously activated via collaterals. In this way, a waking reaction is initiated in the cerebrum. By altering the form of stimulation, the opposite result, a greater or lesser damping, can also be initiated. The reticular formation system is important not only for the distribution of tone and coordination of the musculature but also for the clarity of consciousness via the retrograde association with the cerebrum and for the possibility of improving attentiveness.

6 Improved Performance via Work Patterns, Practice and Training

6.1 Improved Performance through Economic Patterns of Work and Movement

In a performance-oriented society, improvement of performance assumes an important value relative to employment. It is not the place of this book to discuss the advantages and disadvantages of this performance orientation and its causes. The striving toward improved performance can be found in the industrial world and in high-performance sport, although the emphasis is entirely different. The result of performance in the industrial working process is the product or the service. However, we are talking about the direct performance of man, which as a rule is only a portion of the man-machine system. When an improvement in performance is possible in the industrial working process, then it should be done without stressing man to a greater extent. The main point would be to rationalize the man-machine chain, i. e., to attain the greatest possible increase in production with reduced physical or psychic stress on man. This means that the machine is optimally adapted to the physiologic and psychologic requirements. Obviously, practice and training of man also have definite roles. It should be emphasized here that it is not just a question of diminishing the energy requirement of the work, as was done in previous decades with an assembly line system, but also of psychologically and economically arranging the work such that the working place becomes more humane, i. e., satisfaction and interest in work must be considered.

Work physiology touches on some of the aspects of political and industrial economy, as well as on the sciences of work and engineering. The further development of working methods makes it necessary to estimate and calculate where man can appropriately be the working element or where it is better to utilize a machine. If the decision is made that man is indispensable for a working process, then the problem arises of how to correctly use him, based on his anatomical and physiologic characteristics.

Irrespective of humanitarian considerations and considering only the cost of human work, then an example, although incomplete, will make it clear just how uneconomic muscle work is due to the high cost of fueling the human body.

If we assume that 50 sacks with a weight of 100 kg/sack must be transported to a loft which is 10 m high, then a physical work of 50,000 kpm is necessary. The mechanical efficiency of the muscular work can be estimated to be at the most 1–2% because the worker must carry his own body weight each time. For the total work, approximately 10,000 kcal are needed for the working metabolism. If the estimate is that 1,000 kcal of fuel will cost only $1, then the total energy cost is $10.

However, if an electrical sack elevator is used which has a mechanical efficiency of 70%, then only about 0.2 kWh will be required for the 50,000 kpm (mechanical-electric heat equivalent of 1 kpm = 2.72×10^{-6} kWh). At a cost per kWh of 10 cents, the operating cost would then be only 2 cents.

Not only the operating cost must go into this equation, but also the initial outlay and the maintenance costs of the machine; the more complicated the performance of the machine must be, then the more expensive these costs will be. Thus, man must be appropriately utilized where complicated tasks with small energy requirements are needed. At the same time, one is also fulfilling the humanitarian requirements of making working life more interesting and varied.

High levels of stress are not always avoidable. To increase the performance of the worker while using the same stress and intensity, an appropriate adaptation of the machine to the physiology of man is needed. This can be successfully accomplished only with the cooperation of manufacturers and physiologists. The basic principles of this adaptation include avoidance of unproductive movement or lifting of one's own body weight, smaller proportions of static exercise, relief of different muscle groups and reasonable protection against extreme thermal influences.

The placement of pauses during work which exceeds the threshold for prolonged work is also important. As seen on page 250, the course of recovery proceeds exponentially, i. e., the first seconds of the pause are more effective for recovery than the later portion. Therefore, it is much more appropriate to insert many short pauses than one long pause.

Investigations at the work site have shown that an effective rationalization can be reached without increasing the stress on the worker by consideration of these and other factors. The problems today are solvable only with interdisciplinary cooperation; they are being treated by the discipline of ergonomics (work science).

In maximal-performance sports, the primary emphasis is on the improvement of physical performance capacity, which is attained by goal-oriented training. Of secondary importance is the optimal adaptation of the apparatus to the form of movement. It is naturally important whether a cyclist enters competition with the incorrect saddle height, incorrect number of pedal revolutions or is in a position which will cause chafing. In many types of sports, the apparatus is prescribed or standardized in such a way that little latitude remains. Nevertheless,

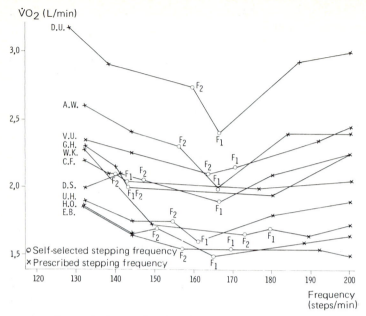

Fig 159 Oxygen intake as a function of stepping frequency at constant speed of 130 m/minute.

the problem will be illustrated by a couple of examples: a "natural" movement form (running) and a sport that is influenced by the apparatus (cycling). It is easy to imagine that an individual can run with three different running styles, namely, short steps at high frequency, moderate stride and reduced frequency or long steps and low frequency. This can be done such that the desired speed is the same in all cases. The question arises as to which relationship between stride length and stride frequency is the most economic. In other words, at what ratio between the two variables does a runner have the lowest O_2 consumption for the work performed?

The results of such an investigation are shown in Figure 159. Stride frequency is shown on the abscissa and the oxygen consumption is on the ordinate. The prescribed velocity for this investigation was 130 m/minute on the treadmill shown in Figure 43; stride frequency was prescribed using a metronome. Almost all subjects had the lowest energy level at 150 to 180 steps per minute. Before and after this series of studies, each subject had the opportunity to determine for himself the frequency (F_1 and F_2) at the prescribed speed which was most comfortable for him and which was spontaneously chosen for the running test. Most subjects ran at the frequency requiring the lowest values of O_2 consumption.

Thus, subjects intuitively select the stride frequencies which are particularly economic. More detailed analysis revealed that the choice of frequency was correlated with the natural frequency of the leg. The leg can be considered the same as a physical pendulum, the natural frequency of which is basically determined by the distance between the center of gravity and the fulcrum. If this distance is altered, for example, by a variable weight of shoe, then man will involuntarily adapt to a new optimal stride frequency and will achieve a minimum level of O_2 consumption at the new frequency.

In cycling there is an opposite pattern. As can be seen in Figure 46 on page 73, there is a clear minimum of energy metabolism for the same cycling performance; this is found in the range of 40–60 pedal rpm. In road racing, the cyclist has the possibility of freely choosing the number of pedal revolutions with the help of a change in gear. Although a fixed gear is prescribed during track racing, it can be adjusted within a certain range before the race. When observing cycling races, it can be seen that most racers select 100–125 pedal rpm, even though the energy cost for the same work is about 1.5–2 times greater than that at 60 rpm (Fig 46). As a result, the anaerobic reserves are taxed to a much greater extent because the threshold for prolonged work, which depends on the maximal local aerobic metabolism, is significantly reduced.

Why do cyclists ride with a pedal frequency which is energetically the least favorable? Studies have shown that the basis for this can be found in a well-known law of sensory physiology. As early as the last century, psychophysical investigations indicated that perception is proportional to the logarithm of the stimulus strength, at least within the midrange. This finding was named after its discoverers and is called the Weber-Fechner law.

In addition to such well-known sensory qualities as seeing and hearing, man also has a sense of tension based on the feedback from muscle spindles and Golgi organs. This sense is also informed by the touch and pain receptors of the skin. If a weight is placed on the hand of an extended arm belonging to a man whose eyes are closed, he can estimate the weight. If the weight and thus the strength of stimulation is doubled, the man will now overestimate the weight and consider it to be more than twice as heavy, while a lighter weight will be underestimated.

Figure 160 shows the relationships between tension and tension perception found for 3 subjects. Tension perception increases more than

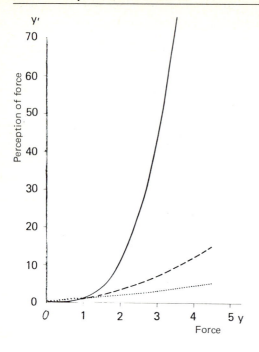

Fig 160
Relationship between force and perception of force in 3 subjects (see text for details).

the tension, although to different degrees in the 3 subjects. The relationship can be appropriately expressed as a power function:

$$y' = y\,\frac{1}{b},\qquad (74)$$

where y' is the perception of tension and y is the tension. In the study shown in Figure 160, b had values ranging between 0.91 and 0.29. Power is the product of force times velocity. In terms of physical power, it matters little how both factors are changed if the product remains constant. From a physiologic viewpoint, a performance composed of a low velocity and a great force is perceived to be more strenuous than a performance in which force is small and velocity high, even when the product of both factors remains constant. Relative to mechanical efficiency, the velocity of 40–60 rpm is optimal. Therefore, the frequency for which the cyclist searches is determined not by the most economic velocity but by the perception of tension.

The perception of tension does not appear to substantially affect the work produced over a definite time period. This can be seen in Figure 161, where a road race was simulated on the bicycle ergometer. The subjects had to produce the greatest possible amount of work (performance × time) over the time of the study; the rate of pedal

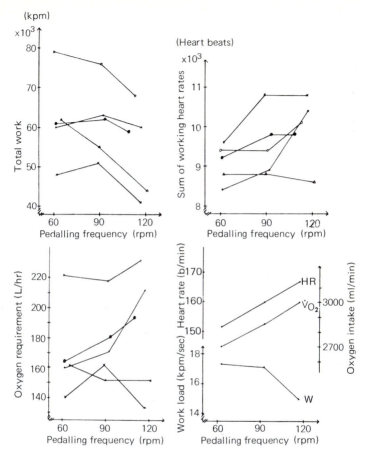

Fig 161   Total work and corresponding physiologic parameters during 1-hour exercise period on bicycle ergometer by cyclists who had task of performing greatest total amount of work at predetermined pedalling frequency. From Ulmer, H. V.: Zur Methodik, Standardisierung und Auswertung von Tests für die Prüfung der körperlichen Leistungsfähigkeit. (Löwenich: Deutscher Ärzteverlag, 1975).

revolutions was prescribed, whereas the subject could adjust the braking force himself. The total work was essentially the same under the three frequencies studied, although the physiologic variables demonstrated that the aerobic portion was greater and the anaerobic portion smaller during the higher pedal frequencies.

The second of the two examples essentially shows that the optimal pattern of work and movement in sport primarily becomes a problem when psychophysiologic influences affect the movement pattern.

6.2 Improvement of Performance Capacity through Practice

Practice is defined as an activity in which improvement in performance capacity is obtained without perceptible organic alteration. As a rule, the gain in performance capacity through practice is greater when the work to be practiced or the athletic movement pattern is more complicated.

The mechanisms which form the basis of motor learning were already discussed on page 39. Practice (the continuous repetition of a movement pattern), forms new, so-called conditioned reactions which essentially produce an automatization of this pattern. This process has also been called conditioning. Simultaneously, condition reactions will be formed which support the movement pattern. It is a common experience that there are a number of cyclists who no longer have a resting heart rate in the laboratory if they see a bicycle ergometer. Just the fact that they have seen the bicycle causes their heart rate to rise. Another example is the rise in heart rate and respiration due to anticipation before the onset of a work load. In this way, the time required for the initiation of the regulatory processes is shortened.
Programs which were once learned are hardly forgotten, even after a long period of nonuse. Although they seem to have disappeared completely, a psychic shock can cause a complete reinstatement of the reaction. In general, automobile drivers are once more capable of driving after a brief practice, even when they have not driven for many years.

This "non-forgetting" of conditioned reactions also has its disadvantages in modern industry, especially if the method of servicing work elements has to be altered. As mentioned earlier, it is inappropriate in a car with automatic transmission to take care of the brake and the gas pedals with the same foot. It would certainly be much better to use the left foot for the brakes and the right foot for the gas pedal because the transfer time would be eliminated in case of danger. The fact that the earlier conditioned reflex breaks through in an emergency, however, could produce false reactions with serious consequences.

The coach or physical education teacher thus has a special responsibility because he brings the first movement pattern of a complicated type of sport to the student. If a false, ineffective technique of the arm pull

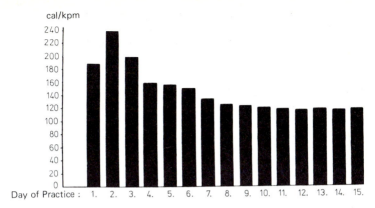

Fig 162 Influence of practice on energy expenditure relative to work done. After first days of practice, there is reduction in expenditure; this denotes improved coordination. From Lehmann, G.: Praktische Arbeitsphysiologie. (Stuttgart: Thieme, 1962).

was learned in the breast stroke, then this movement stereotype would be formed. If the student then wished to become a competitive swimmer, he would find that relearning would be more difficult than learning from scratch. It is also possible that under the stress of competition, the old stereotype would take effect once more. In the case of such technically difficult types of sports as skiing, lessons should be taken from the beginning. A self-taught, false movement pattern can be a great hindrance on the way to perfection.

Due to conditioned reactions, acquired and innate behavior patterns can come into conflict. In skiing, the innate behavior pattern tempts the beginner to put his weight on the uphill ski while traversing and to lean into the hill so that he will not have so far to fall if he tumbles. The beginner will be instructed that he should put his weight on the downhill ski because of the more favorable position of his center of gravity. He must shift his weight downhill. Although the beginner will do the right thing when there is no danger, when there is a chance of falling he will react incorrectly and thus fall.

The effect of the conditioned reaction on coordination is such that the movement pattern will become more and more economic. The economization occurs primarily because the superfluous collateral movement of muscles which do not belong to the movement pattern is reduced, thus sparing energy. The tension of the antagonists and that of the agonists will also adapt, so that no superfluous tension will be developed. Figure 162 shows the reduction in energy metabolism for an activity which is practiced daily. The columns representing the first days of practice show the economy ratios (see p 71), which are a measure of the mechanical efficiency for the condition before practice. At first, there is a rise in energy expenditure on the 2nd and 3rd day. This is probably associated with the effect of muscle soreness. The oxygen requirement becomes continuously smaller from the 3rd to the 9th day and then remains constant.

In the world of sports, practice (learning techniques of a type of sport) and training (increase in endurance or strength) usually occur in parallel. The question of reciprocal effects will be found in the discussion. Many well-known trainers believe that interval training, during which significantly more lactic acid appears than during endurance training, is less favorable for the formation of coordination, because acidosis supposedly inhibits the formation of conditioned reflexes. Because the neurophysiologic bases of the formation of conditioned reflexes are not yet clear, this theoretical question is unanswered.

6.3 Increased Performance Capacity through Training

Training is the effort to maintain or to improve physical performance capacity over a prolonged period of time through goal-oriented physical activity. The Roux rule is valid here: small stimuli are useless, average stimuli are useful and large stimuli are harmful. In general, the performance capacity that is trained will be increased. Strength training increases muscle strength, whereas endurance training improves only endurance capacity. Physical inactivity causes a reduction in performance capacity. Thus, adaptation occurs in both the positive and negative direction.

The working world in the industrial states today is characterized by those occupations with physical inactivity (office work) or repetitious work demands (assembly line work). On the other hand, there is an excessive activation of the adreno-sympathetic system due to environmental stresses (noise), psychologic stress (time pressure) or unhealthy leisure time demands (long weekend drives in the car). These and other stress factors, associated with the absolute or relative lack of movement, produce the well-known risk factors for the cardiovascular system; these risk factors can be partially offset by vagotonia, such as is obtained with endurance training.

In the view of sports physiology, the special importance of training is that it helps to obtain a certain level of performance. In so doing, however, it can be presumed that many areas of competitive sports greatly surpass the usefulness needed for good health. Training primarily improves performance capacity by measurable changes in organs and organ systems, from which technique, style and tactics for the individual types of sports are then formed through practice. This allows the athlete to attain the most economic and most effective movement pattern, so that the fitness which has been obtained can be exploited fully.

The fact that famous trainers or even schools of sports scientists often have bitter feuds and consider their training methods as the only true way to train shows that even in the physiology of training there are still many unanswered questions. It also shows that knowledge alone is not sufficient to be a good trainer. The physiologic science of training can be only a part of a complex operation that must also include psychologic and sociologic aspects. Without the presence of objective criteria, an additional amount of appropriate weight must be given to intuition and personal experience.

The future task of training science will be to obtain more objective data. Data obtained from many aspects of high-performance sports can then be usefully applied to mass sports and rehabilitation.

6.3.1 Principles of Long-Term Adaptation of Performance Capacity

Certain basic principles apply to the adaptation to the required endurance, as well as to the required strength. These principles primarily relate to the relationship between stimulus and alteration of the organ structure. Each nonutilized organ system has the tendency to lose its capacity to perform. This is observed with time as a continuous atrophy in muscles whose corresponding motor neuron no longer gives any impulses due to disease (infantile paralysis) or injury (paraplegia). As a result, the muscles initially become thinner and lose their strength. In its final state, muscle tissue shrivels up entirely and is transformed into connective tissue. This damage is irreparable. Even if the function of the motor neurons returned, there would be no retransformation in the specific muscle tissue. During this progressive reduction in function, both strength and endurance are lost.

Under normal conditions, the tendency toward a reduction in the ability to function is compensated via stimuli which improve function. The resulting performance capacity of a muscle is characterized by a continuous equilibrium between the natural breakdown in performance capacity and the gain in performance capacity due to adequate functional stimuli, even though they can specifically maintain or improve a definite function. If strength stimuli are applied, then only strength is improved; if endurance stimuli are applied, only endurance increases.

Muscular movement during daily living and the functional stimuli applied cause adjustment of performance capacity to the value corresponding to the average requirement. However, the technical development of our environment has brought with it the fact that the amount of functional stimulus has markedly decreased. It is easy to imagine that it is of little consequence for improving performance capacity whether a person walks up the stairs to the 3rd floor or whether he uses an elevator. In all areas of life today, physical activity has been reduced and there is a definite tendency toward comfort. It is certain that public enterprise means well when it installs escalators in the subways of major cities. However, it would be much better if all physiologic training stimuli were not taken away from man.

From the principles described, it can be inferred that performance capacity can adapt to higher or lower levels and that the greater the difference between the actual state of the performance capacity and the positive or negative training stimuli, the greater the adaptation will be. If a completely untrained person applies an intense but constant training stimulus, then the increase in performance capacity at the onset will be much greater than that shortly before the new steady state is attained. In the same sense, if a well-trained person suddenly stops training, he loses considerably more work capacity in the first few days than he will lose in the period that follows. As a first approximation, the increase and decrease in performance capacity has the form of a negative exponential function.

The concepts about the training effect and its initiating mechanism are for the most part scanty. There are numerous studies which have looked at the final state in the form of cross-sectional investigations and in which data of an untrained group are compared to those of a trained group. However, longitudinal studies, especially on endurance training, which systematically follow the course of training, are seldom seen. The physiologic bases of the problems which have been noted for the longest time are known; these problems appear with a sudden increase in training and the sudden end of training of a highly-trained person. With an increase in training, "overtraining" can sometimes occur. This phenomenon is especially impressive with endurance training. The athlete feels distressed and his performance capacity suddenly drops off. At the same time, such objective symptoms as ECG disturbances are found. They are probably evoked because the different links in the chain that increase performance capacity as an integral process adapt at different rates to the increasing performance. With interval training, for example, the strength of the heart is trained more than the endurance of the heart, because it produces more of a pressure elevation. During marathon training, it is estimated that the endurance of the heart is increased more through the increase in mitochondrial enzymes. If strength adapts faster than endurance, there is obviously a point where the heart undergoes a relative oxygen deficiency during stress because it can mechanically perform more than it can cope with metabolically. In this condition, the T wave of the ECG will be flat for several days, even at rest; this is similar to occurrences under pathologic conditions with coronary insufficiency. If the training is reduced, the condition will rapidly return to normal.

Even the autonomic adjustment of cardiac output and blood flow or of blood flow and ventilation apparently are closely adapted to one another. Sudden increases in training can disturb the harmony of these control systems to such an extent that pathologic symptoms appear.

If there is a sudden stop in activity in a well-trained athlete, either due to an accident or due to the fact that the athlete ends his career, similar problems appear for the same reasons. For this reason, even well-trained athletes should not stop training abruptly, but should slowly reduce their daily training.

6.3.2 Fundamentals of Isometric Strength Training

The adequate stimulus for an increase in muscle strength is the tension of the muscle fiber; this tension must be so strong that it surpasses the stimulus threshold. The greater the difference from the actual performance capacity, the greater the training stimulus will be. When the fastest and most effective increase in strength is desired, the greatest possible training stimulus must be applied; the athlete must exert his muscle with its maximal tension. The force of a muscle during a static (isometric) contraction is always larger than that during a dynamic (auxotonic or isotonic) contraction.

To determine the natural laws, we will first consider a study on the influence of maximal isometric contractions on muscle strength. Subjects had the task of stressing a muscle with its maximal force for one second daily. Maximal strength increased regularly so that the load each day became somewhat greater. This method is called "progressive training". As shown in Figure 163, maximal strength increased daily and after about 5 weeks approached its final value in an asymptotic fashion. The subjects studied behaved in approximately the same manner. Nevertheless, such a graph is not without problems because it always assumes that the muscle groups studied operate under essentially the same initial conditions. Because some muscle groups are stressed more than others in daily life, there are also large differences in the initial strength of the different muscle groups and differences in the increase with progressive training. In physically untrained persons, muscles of the body which are often used (for example, the forearm flexors or extensors) already have 76–80% of the final level of strength which can be obtained with a training program of daily maximal contractions of 1 second. For these reasons, data relating to trainability are not uniform.

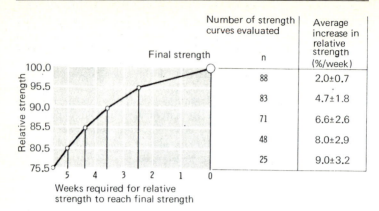

	Number of strength curves evaluated	Average increase in relative strength (%/week)
	n	
	88	2.0±0.7
	83	4.7±1.8
	71	6.6±2.6
	48	8.0±2.9
	25	9.0±3.2

Fig 163   Increase in muscle strength with progressive training as a function of time. From Müller, E. A. and Rohmert, W.: Int. Z. angew. Physiol. 19 : 403, 1963.

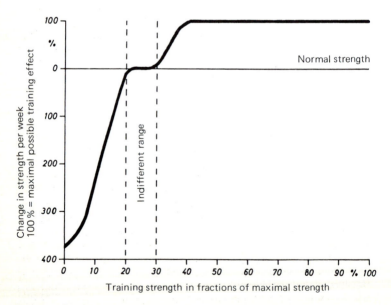

Fig 164   Changes in strength with training in relation to training strength. From Hettinger, T.: Isometrisches Muskeltraining, 4th ed. (Stuttgart: Thieme, 1972).

Whether the final level of strength shown in the studies actually exists or whether further increases in strength can result from the same type of training after a maintenance period and further weeks of training is controversial. It is likely that during prolonged strength training, muscle strength can be further increased, even under the same conditions. The greater the difference between the actual and the final level of strength, the greater the increase in strength will be. The final level of strength is a constant value for only one type of training. It can be increased further if the requirement is raised, e. g., if the time that the muscle is stressed increases from 1 to 5 seconds. Once again, muscle strength increases in the same way asymptotically to a new final strength.

Systematic training studies have produced the surprising fact that muscle strength can be increased with little effort and cost (Fig 164). On the abscissa is training strength in fractions of maximal strength which the muscle can produce. The experimental conditions in this study were selected such that the muscle being studied was placed in the most relaxed situation possible and then exerted only once a day against a force requiring a percentage of the maximal strength. If the tension was 20–30% of the maximal strength, then muscle strength remained the same. If it was less than that, the muscle had an acute loss of strength (muscle atrophy). Between 30 and 50% of the maximal strength, there was a marked rise in muscle strength. The optimal number of training stimuli was 3–5 isometric contractions per day of more than 60% of the maximal force. Several stimuli of this type did not permit muscle strength to increase any further.

Because of the practical importance of this laboratory investigation for strength sports, the results have received a great amount of attention and discussion relative to their practical application for these sports. Strength athletes repeatedly indicate that they have had the personal experience of achieving better success in terms of an increase in strength by training with dumbbells or on a strength machine with dynamic strength exertion. As a result, the suspicion arises that the results may be false. To avoid misunderstanding, some remarks must be made. For reproducible results, the muscle group to be trained should be trained at a constant angle and under identical conditions. At the most, only a portion of a muscle group or of a muscle will be tested; this depends primarily on anatomical factors. The results may be considered valid only for the muscle fibers which are stressed at the given length. A different angular adjustment stresses other muscle fibers and muscle groups which have not been trained. With dumbbell training or training on a strength machine, there will be quasi-isometric contractions of different fibers or muscle groups during slow movement, such that strength is trained at all of the angles used.

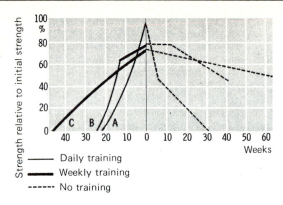

Fig 165   Pattern with training and stability of changes induced by training. From Müller, E. A. and Hettinger, T.: Arbeitsphysiologie 15 : 15, 1954.

Questions about the conditions which initiate an increase in muscle strength are still unanswered. Animal experiments have shown a clear increase in fiber diameter, such that the myofibrils become thicker and more numerous. The muscle thus becomes thicker and stronger due to hypertrophy. There are occasional reports of hyperplasia, which is an increase in the number of muscle fibers. Whether this also occurs in man is not known at this time.

The absolute voluntary muscle force per unit of cross section is about 6 kp/cm^2 under physiologic conditions. This remains constant even during strength training because strength is a linear function of the cross section. Normally, only about two-thirds of all muscle fibers can be voluntarily innervated at the same time. During electric stimulation of all fibers, an absolute strength of about 10 kp/cm^2 is obtained. Of course, the cross section must be measured perpendicular to the direction of the muscle fiber; this is especially important in pinnate muscles.

A muscle which is brought to a higher final level of strength with isometric strength training is specialized in strength work. The ability of such a muscle to work dynamically generally decreases. This is because during the brief isometric exertion, only a stimulus for the increase in fiber thickness is applied. With inactivity, muscle strength gained by training is quickly lost again. Figure 165 shows that the time in which the strength was gained is of primary importance relative to how fast this strength will be lost again. Strength which is gained rapidly is lost faster than strength which is gained slowly.

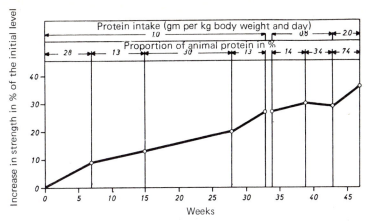

Fig 166   Increased strength with training once a week in relation to protein intake. From Kraut, Müller, E. A., Müller-Wecker: in Hettinger, T.: Isometrisches Muskeltraining, 4th ed. (Stuttgart: Thieme, 1972).

6.3.2.1 Physiologic and Pharmacologic Influences on Strength Training

The increase in strength with isometric training is associated with the synthesis of contractile protein (p 7). Due to the DNA-RNA system, the individual amino acids are put together in a series of building blocks to synthesize specific proteins; in muscle, the proteins are actin and myosin. Proteins are taken up in the diet and broken down in the digestive process. If the necessary proteins are not present in the diet, then no synthesis of muscle tissue can occur.

Untrained persons normally have a daily protein requirement of 1 gm/kg body weight. If a training effect on large muscle groups is desired, then this amount must be increased. The results from a study shown in Figure 166 reveal that an increase in muscle strength with progressive training can be obtained with a protein intake of 1 gm/kg body weight. During the first 33 weeks, only the proportion of animal protein was varied. The problem being investigated was whether the proportion of essential amino acids (p 80) known to be higher in animal protein would affect the increase in muscle strength to any degree. An exact analysis showed that this was not the case. With continuation of training and an intake of 0.8 gm/kg body weight, the increase in strength leveled off, whereas it was further increased with 2 gm/kg body weight. No increase in strength is possible without a surplus of protein.

The synthesis and rate of degradation of proteins is controlled by the androgens (male sexual hormones). For these reasons, the trainability of muscles is closely associated with the concentration of testosterone, the most important androgenic agent. Because androgens which have been slightly chemically altered are frequently used to obtain anabolic effects with strength training (this is naturally included under the term "doping"), the physiologic regulation of testosterone levels and its effects will now be discussed in detail.

Testosterone is chemically a C-19 steroid with an OH-group at C 17. It is synthesized from cholesterol in the Leydig interstitial cells of the testes. The testosterone secretion in the normal man is 4–9 mg/day. It is controlled by the luteinizing hormone (LH) of the anterior pituitary. Women also form small amounts of testosterone, possibly in the ovaries. The hormone is also synthesized to a small extent in the adrenal gland of both sexes.

Approximately 60–70% of plasma testosterone is bound to protein. Its concentration (free and bound) in young men is 0.6 µg/100 ml, with the values varying between 0.5 and 0.8. The normal woman has 0.1 µg/100 ml. Testosterone concentration is reduced about 20–30% with increasing age. Before puberty, both sexes have approximately the same concentration, which then increases about tenfold during puberty in boys. As a result, the secondary sex characteristics are formed. A large proportion of testosterone is continually converted in the liver to 17-ketosteroids and eliminated in the urine. These also have a weak androgenic effect. Production and degradation occur relatively quickly, with half of the testosterone formed being broken down in about 4 minutes.

The concentration of testosterone is kept constant via a servocontrol loop (Suppl, p 303), such as is illustrated in Figure 167. The sensor of the control loop is located in the hypothalamus. If, for example, testosterone concentration falls, two substances will be increasingly secreted: the releasing factor for LH (LRF) and for follicle-stimulating hormone (FSHRF). Both factors cause an increase in the release of FSH and LH from the hypophysis. As long as LH is elevated, the Leydig interstitial cells will be informed via the blood stream and will then excrete more testosterone. At the same time, the formation of sperm is promoted by FSH using the same pathway. Too much testosterone produces the opposite reaction. Animal experiments have shown that implanting the smallest amount of testosterone in the hypothalamus inhibits spermatogenesis so strongly that the seminiferous tubules atrophy; if this continues for a prolonged time, the sperm-forming apparatus will become functionally inactive, resulting in infertility.

The trainability of muscle, therefore, depends greatly on testosterone concentration. The main result of all of this is the difference in

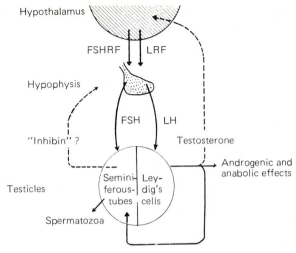

Fig 167   Postulated relationship among hypothalamus, anterior hypophysis and testicles. Stimulating effect, solid arrow; inhibiting effect, dashed arrow. From Ganong, W. F.: Medizinische Physiologie. (Berlin: Springer, 1971).

trainability as a function of age and sex. Figure 168 shows the trainability in per cent of maximal trainability for both sexes as a function of age. There are little differences before puberty; after puberty these differences are marked, only to become similar once more in old age.

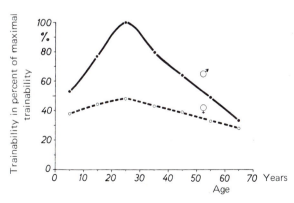

Fig 168   Trainability of muscles in extremities in relation to age and sex. From Hettinger, Th.: Isometrisches Muskeltraining, 4th ed. (Stuttgart: Thieme, 1972).

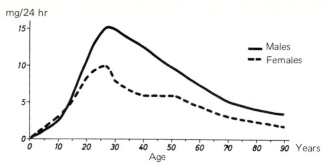

Fig 169   17-ketosteroid excretion in relation to age and sex. From Hettinger, T.: Isometrisches Muskeltraining, 4th ed (Stuttgart: Thieme, 1972).

The excretion of 17-ketosteroids, which is a measure of testosterone production, is shown in Figure 169. When the two figures are compared, the similarities are apparent.

For a long time, the pharmaceutical industry has been developing substances with effects similar to those of testosterone. These substances are collectively termed "anabolic steroids." They are used in medicine as therapy for protein deficiency problems after severe infections or due to metabolic disturbances or malignant tumors. Unfortunately, these substances are frequently used in sport, contrary to doping regulations, in order to support strength training and to make it more effective. Anabolic steroids are chemically similar to the structural formula of testosterone, with slight modifications of the molecule. For example, the transposition of hydroxyl or methyl groups on the C-19 steroid diminishes the sex-specific component and amplifies the anabolic components. From a medical standpoint, there are many reasons for warning against the abuse of anabolic steroids in sport. It is certain that an uncontrolled interference with hormone levels is not as harmless as many athletes believe and it has not yet been clearly determined whether permanent sterility can occur as a delayed consequence. In addition, the effect of testosterone is closely associated with psychic reactions which are not always advantageous. Finally, a much more important point remains to be mentioned, i. e., the danger of unpleasant sports injuries due to pulled tendons, torn ligaments and injuries to the bones. Tendons, ligaments and bony structures are basically trainable, but this occurs much more slowly than with muscle strength. As a rule, the trainability of a tissue is proportional to its metabolic level at rest. If the increase in strength is too rapid due to the support of anabolic steroids, the other tissues mentioned above do not have a chance to adapt to the stress in a similar manner.

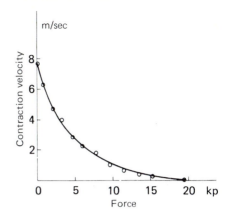

Fig 170
Relationship between muscle contraction velocity and load.

6.3.3 Fundamentals of Training for Speed

Speed can be improved by strength training and by improving coordination with practice. The greater the stress on an individual muscle, the slower will be the contractile velocity of that muscle (Fig 170). Consequently, the absolute contractile velocity depends on the reserve strength, i. e., the difference between maximal force and the actual force applied. For the same externally delivered force, reserve force is of primary importance for the velocity of contraction.

The initial acceleration that occurs with the transfer of the body is decisive for the speed in an athletic activity, e. g., 100 m sprint. Because force = mass × acceleration, the effect of force is evident.

6.3.4 Fundamentals of Training for Endurance

Local muscular endurance is limited by the relationship between aerobic and anaerobic energy supply (p 49). To obtain a training effect, it is necessary that the muscle be stressed for the longest possible time and in such a way that it just surpasses the limit of the aerobic energy supply. This can be done in two ways. Either the muscle is stressed so that the threshold of prolonged work is just barely surpassed (the muscle is stressed for a relatively long time before it fatigues) or the muscle is trained with varying intensity, during which the fatigue phase and the recovery phase are alternated.

Only a few years ago, such improvements in endurance were thought to be due to only one factor, capillarization. Therefore, a great deal of interest was devoted to this factor in older studies on endurance training. We know today that this is not as important as once imagined. According to the current view, the much more crucial point is the adaptation of the cellular metabolic processes due to the increased formation of enzymes in muscle mitochondria.

6.3.4.1 Effect of Endurance Training on Cellular Function and Structure of Muscle

Systematic endurance training studies to explain cellular adaptation were first carried out on rats. Recent investigations done with man have shown approximately the same results. The difficulty with human studies is that a 6-mm thick needle must be inserted into the muscle for each investigation to take a muscle biopsy specimen. It is not easy to obtain representative fiber groups because the number of muscle samples obtainable is understandably limited. Therefore, animal research can often give much more complete information.

In rabbits, for example, it is easy to distinguish between the "fast" (phasic, white) muscle fibers and the "slow" (tonic, red) fibers. The fast fibers are especially suited for anaerobic performance and the slow fibers are predominantly suited for endurance performance. In rabbits, there are muscles which are predominantly composed of white fibers and other muscles which are predominantly composed of red fibers. In man, this differentiation is found within the same muscle. Red and white muscle fibers often lie adjacent to one another.

Table 20 Examples of Predominantly Phasic or Tonic Muscles in Man. From Keul, J. et al.: Muskelstoffwechsel (Munich: Barth, 1969).

Synergistic muscle group	Phasic	Tonic
Quadriceps femoris	vastus internus and externus	rectus femoris
Triceps surae	gastrocnemius internus and externus	soleus
Flexor of the knee joint	semimembranosus	semitendinosus
Triceps brachii	caput ulnare and radiale	caput longum
Fixators of the shoulder joint	latissimus dorsi	infraspinatus
Other muscle		rectus abdominis sacrospinalis
	tibialis anterior biceps brachii adductor digiti V deltoideus flexor digitorum profundus	

Table 20 gives information about the distribution of red and white fibers in different muscles. This classification is primarily valid for untrained persons because it has been shown that white muscle fibers can be converted into red fibers with endurance training.

Innervation has a major role in the primary formation of red and white muscle fibers; these fibers are already distinguishable in the embryonic state. In the case of red fibers, a larger (trophic) discharge frequency is produced over the neuromuscular synapse, possibly from the corresponding motor neurons.

Denervated muscles show a transition from red to white fibers, whereas prolonged stress has the opposite effect.

Table 21 contains a summary of several enzymes which control the various energy supply systems. Careful analysis indicates that glycolytic activity (the ability to supply energy anaerobically) is much greater in white muscle. Obviously, white muscle can also mobilize glycogen better in the cell. Because of their enzyme content, red fibers are particularly suited to supply oxidative energy, as can be seen from the markedly higher enzyme activity for the citric acid (citrate) cycle and the respiratory chain. Red fibers have particularly developed the ability to draw on fatty acids due to their enzymatic supply; this is especially important for prolonged exercise.

From a mechanical standpoint, white muscle has a shorter contraction time than red muscle. It also has a higher membrane potential and a stronger excitability. Therefore, it is particularly well-suited for the development of power.

To investigate the effect of endurance training on enzyme activity, a study was carried out in which rats performed exhausting swimming training, as shown in Table 22. To be certain that the training reached the necessary intensity, the animals were also progressively stressed with weights. The results demonstrate that the enzymes for glycolysis were changed only a small amount. Glycogen in the muscle increased to 150% of the control value. The enzyme activity levels of 5 enzymes used in aerobic metabolism were also studied (Fig 171). Cytochrome a, b and c values increased to about 140%, when their level was related to the amount of mitochondrial proteins. Succinate dehydrogenase (SDH), which was also related to the mitochondrial protein content, increased at first and then essentially reverted to the control value. The same pattern was found for glycerol-1-phosphate oxidase (GP-OX), which even dropped to about 20% less than the control value at the end of the study. However, Figure 172 shows that the total content of mitochondrial protein increased markedly from the 12th week. Collectively, there was a considerable elevation of all the values studied. Electron microscopic studies also showed that the number of mitochondria increased and that the mitochondrial cristae were mar-

Table 21   Comparison of Various Enzyme Activities in White (Phasic) and Red (Tonic) Muscles (from Keul, J., et al.: Muskelstoffwechsel [Munich: Barth, 1969]).

Enzyme	Unit of measurement	Phasic muscle	Tonic muscle	Species
	Degradation of carbohydrate			
Phosphorylase	μm/gm/hr	2,800	135	rabbit
Phosphorlylase	μm P_i/gm/min	32	7.5	white rat
Hexokinase	μm/gm/hr	12	93	rabbit
Hexokinase	units/mg/protein	4.64	5.08	guinea pig
Aldolase	units/gm/noncollagen protein	1,230	96	cat
Aldolase	units/gm	320	32	white rat
Triosephosphate isomerase	μm/gm/hr	6×10^5	8×10^4	rabbit
Glyceraldehyde phosphate dehydrogenase	μm/gm/hr	87,260	10,850	rabbit
Enolase	μm/gm/hr	10^4	10^3	rabbit
Pyruvate kinase	units/gm/noncollagenprotein	1,620	270	cat
Pyruvate kinase	units/gm	850	182	white rat
Lactate dehydrogenase (total)	μm/gm/hr	218.4	114.5	man
Lactate dehydrogenase (total)	units/gm	52,625 2.00	750	white rat
Hexose diphosphatase	units/gm	350	0.255	rabbit, chicken
α-Glycerolphosphate dehydrogenase	units/gm	10	22	white rat
Glucose-6-phosphate dehydrogenase	units/kg noncollagen protein		30	rabbit
Glucose-6-phosphate dehydrogenase	units/gm	0.2	0.42	white rat
Phosphoenolpyruvate carboxy kinase	mU/gm	3.3	6.5	guinea pig
	Degradation of fats			
Hydroxyacyl-CoA dehydrogenase	μm/gm/hr	62	250	rabbit
	Degradation of amino acids			
Glutamate oxalacetate transaminase	μm/gm/min	35.8	42.8	man
Glutamate oxalacetate transaminase	units/gm	27	94	white rat

Table 21   (continued)

Enzyme	Unit of measurement	Phasic muscle	Tonic muscle	Species
Glutamate pyruvate transaminase	μm/gm/min	4.2	6.3	man
Glutamate dehydrogenase	μm/gm/min	0.98	0.24	man
Glutamate dehydrogenase	μm/gm/h	18	40	rabbit

Citrate cycle

Condensing enzyme (citrate synthase)	μm/gm/hr	86	340	rabbit
Isocitrate dehydrogenase	units/gm/noncollagen protein	355	855	cat
Isocitrate dehydrogenase	units/gm	5	41	white rat
Fumarase	μm/gm/hr	86	340	rabbit
NAD-Malate dehydrogenase	μm/gm/min	240	335	man
NADP-Malate dehydrogenase	mU/gm	85	137	guinea pig

Conversion of energy-rich phosphate

Creatine kinase	mgP$_i$/gm/hr	37	34	rat
Creatine kinase	μm/gm/min	196	126.4	man
ATPase	μm/mg/hr	199.4	59.5	rabbit
Ca^{++}-activated ATPase	μmP$_i$/mg myosin · min	0.34	0.14	rabbit
Adenylate kinase	units/gm	60	296	rabbit
Cytochrome oxidase	units/gm	2	4.2	white rat
NADH-oxidase system	μm/gN/min	134	279	guinea pig
NAD-NADP trans-hydrogenase	μm/gN/min	11.5	19.9	guinea pig

Table 22   Effect of Training on Swimming Performance of Rats (H. Kraus, R. Kirsten, J. R. Wolff).

Swimming Duration (Weeks)	Number of Animals	Weight Attached gm/100 gm Body Weight	Average Daily Swimming Time (min)
2	12	3.5	16.7
4	12	4.5	24.7
8	12	4.9	32.8
12	9	5.3	39.3
15	9	5.9	41.5

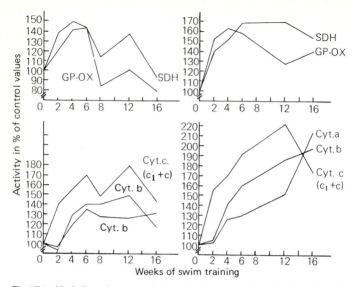

Fig 171 Variations in several important enzymes in mitochondria due to training rats by swimming, presented as per cent difference from corresponding values of control animals. From Kraus, H., Kirsten, R., Wolff, J. R.: Die Wirkung von Schwimm- und Lauftraining auf die zelluläre Funktion und Struktur des Muskels. Pflügers Arch. 308 : 57–59, 1969.

kedly thicker. The increased activity is obviously due to a real synthesis via enzyme induction.

Essentially the same results have been found in man. Endurance training is evidently capable of increasing not only the oxidative capacity of the muscles but also of converting white muscle fibers into red muscle fibers. Quantitative investigations about the extent of this conversion are lacking in man. From a physiologic viewpoint, it cannot be decided whether such training will have negative effects on power development (for example, in a high jumper).

6.3.4.2 Endurance Training and Capillarization

For many years, the effect of capillarization in muscle as a result of endurance training was considered to be the factor which alone determined local muscular endurance. For about 10 years, it has been known that the cellular adaptive processes described in the last section represent a much more decisive factor. The concept that the capillarization of a muscle determines its endurance is attributed to the classic observations on capillaries by the Danish physiologist Krogh, who

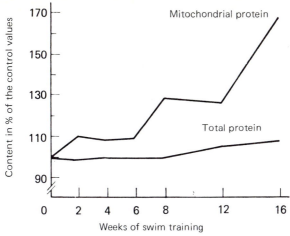

Fig 172   Changes in total and mitochondrial proteins during swim training of rats. From Kraus, H., Kirsten, R., Wolff, J. R.: Die Wirkung von Schwimm- und Lauftraining auf die zelluläre Funktion und Struktur des Muskels. Pflügers Arch. 308 : 57–59, 1969.

showed that there are muscles with varying levels of capillary density. It is now known that only a portion of the capillaries are perfused at rest; the others contain no blood. During exercise, peripheral resistance in the working muscles is reduced because the afferent smallest arteries are dilated (p 101). As a result, some or all of the capillaries which were empty at rest will be passively dilated, depending on the level of metabolism. In this way, they supply increased blood to the muscle.

A nonperfused capillary can be described as an extremely thin endothelial tube. To count the number of capillaries supplying a muscle fiber, the light microscope is used. To be able to recognize the capillaries, it is necessary for them to be full. In animal experiments, capillaries of the still-living animal are filled with a staining fluid, cross sections of the musculature are made after the animal is killed and the number of capillaries is counted. It has been shown that a much greater number of capillaries can be seen with optimal filling techniques than when there is inadequate filling. Because it is understandable that no filling technique can be used in man just before his death, there are obviously some doubts about results published earlier. As a result, many authors today are no longer seriously convinced that there is actually a real synthesis of capillaries due to endurance training.

A large part of our physiologic knowledge originates from investigations on animals. However, results from animal studies are particularly uncertain, especially regarding this question, because different animal species are specialized for entirely different performances. For example, the dog is a definite running animal. Hence, it follows that relative to its body weight, the dog has a heart which is at least twice the normal size. Thus, if a greater capillary density is found in dogs, this does not imply that the same is needed by man.

On the other hand, there are several kinds of animals in which a real new growth of capillaries due to endurance training can be observed using optimal techniques. Man at least has the ability to form new capillaries due to inflammatory stimuli. The cornea of the eye is normally supplied via diffusion because no capillaries are located there. After inflammation of the cornea due to infection, capillaries grow into the cornea. Analogous to this process, the possibility of a new growth of capillaries is imaginable in muscle.

During the course of effective endurance training, all the symptoms of inflammation can be detected. The classic symptoms of "calor, tumor, rubor and dolor" of inflammation are experienced at times by almost every athlete who has trained hard. Calor (heat) is demonstrated in the form of slight fever and prolonged high temperature in the muscle groups which were stressed. Tumor (swelling) can be seen when the muscle group is palpated. While rubor (reddening) cannot be observed through the skin, it can be objectively measured by a stronger blood flow to the organs. Dolor (pain) is known to all athletes as the phenomenon of muscle soreness.

It is certain that muscle soreness has nothing directly to do with the concentration of lactic acid, as is often heard. If this were the case, it would have to be present a short time after the end of training before lactic acid is removed. The other symptom known to the physician and associated with inflammation can also be determined, i. e., an increase in the number of leukocytes and of certain protein bodies in the blood. This "sterile" inflammation is possibly provoked by the marked acidification and microtrauma of the tissue.

Although it is not certain whether there is a capillarization due to an increase in the number of capillaries per muscle fiber, we will now concern ourselves with the effect of capillarization for the following reasons. The endurance-trained individual reaches a markedly higher maximal cardiac output and has a greater possible range for adjusting sympathetic tone. Both effects mean that working muscle will have a greater maximal perfusion. Whether this is due to more capillaries being formed or whether previously functionally-closed capillaries are perfused is, in the final analysis, of little importance for the mechanism of tissue supply. What is important is that the diffusion pathway is markedly shortened in both cases.

The effect of increased capillarization is due to the fact that flow velocity in the individual capillaries is less with the same pressure

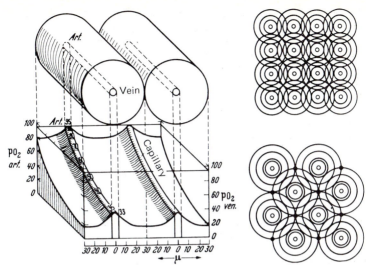

Fig 173   Diagram of tissue oxygen supply as function of capillary length and of O_2 pressure. Oxygen pressure drops markedly in the area immediately below the capillary. Right, shortening of diffusion pathway when there is increased capillarization due to training. From Schneider, M.: Einführung in die Physiologie des Menschen. (Berlin: Springer, 1966).

difference at the entrance to and exit from the capillary bed. In addition, there is an increase in the surface for exchange. Both effects produce a greater ability to release oxygen from the blood, so that the maximal AVD for oxygen increases. According to the Fick diffusion law, the amount of oxygen diffused is proportional to the pressure difference between blood and mitochondria, whereas the diffusion pathway is inversely proportional. A cylinder of tissue (Fig 173) is supplied with oxygen by each capillary. If capillarization is insufficient, there are possibly sites with such a low O_2 pressure, resulting from the poorly supplied region between the tissue cylinders, that a small portion of the energy in that region is supplied anaerobically at rest. In any case, skeletal muscles of untrained persons at rest always release greater amounts of lactic acid than those of trained individuals.

Data on resting blood flow of skeletal muscle is somewhat variable. It lies somewhere between 3–5 ml blood per minute per 100 gm muscle, so that an average value of 4 ml blood per minute per 100 gm muscle can be assumed. In contrast to earlier statements that the resting blood flow in trained persons was lower, in reality it is somewhat higher; this is caused by the lower sympathetic tone and the lower tone of the smallest arteries. Even these findings are disputed. The O_2 require-

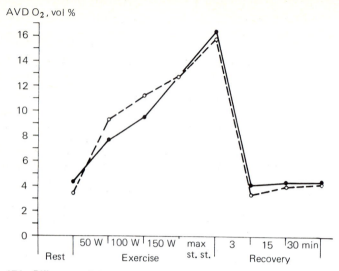

Fig 174   Differences between arterial and femoral venous oxygen content of 20–30-year-old normal persons (open circles) (n, 14) and athletes (closed circles) (n, 12) at rest, during steady-state exercise and in recovery phase. Trained persons reached 300 W at maximal steady-state exercise and untrained persons 200 W. From Keul, J., Doll, E., Keppler, D.: Muskelstoffwechsel, (Munich: Barth, 1969).

ment of resting muscle is about 0.15 ml/minute · 100 gm muscle so that from this there is AVD of about 5 vol% O_2, corresponding to 23% saturation. In this case, the values for trained and untrained persons are about the same.

The effect of endurance training is particularly shown by the level of muscle metabolism which can still be supplied aerobically. The maximal blood flow of untrained persons is approximately 130 ml blood per minute per 100 gm muscle, whereas trained individuals have 180 ml blood per minute per 100 gm muscle, i. e., it is about 40% higher. The O_2 requirement at maximal steady-state exercise is 20 and 35 ml O_2 per minute per 100 gm muscle for untrained and trained individuals, respectively. In both groups, the AVD for O_2 increases, reflecting the higher extraction during maximal exercise.

This fact can be seen in Figure 174, in which AVD for O_2 is expressed as a function of work load. Trained persons extract somewhat less oxygen during submaximal exercise; this is due to the fact that their sympathetic tone is less and their blood flow is somewhat greater. Within the range of maximal steady-state exercise, which is higher in trained persons in terms of absolute work performed, the extraction is approximately the same.

6.3.4.3 Effect of Endurance Training on the Transport Capacity of the Heart

Similar to what has already been described for skeletal muscle, the effect of training on the heart is due to an increase in the force of the individual myocardial fibers and the improvement in their endurance. The heart cycle (p 104) consists partially of isometric contractions and partially of contractions against increasing resistance (auxotonic). Depending on the form of stress on the heart, either that component which produces hypertrophy of the myocardium or that which improves the endurance of the heart will predominate.

Similar to what was seen with skeletal muscles, increased strength of the heart is particularly related to the tension of the fibers with which the heart contracts. We must therefore consider those factors on which the strength of the individual muscle fibers depends.

If the myocardium is imagined in a simplified manner as a hollow sphere (Fig 175), then the approximate force development of the individual muscle fibers can be calculated. There is always an equilibrium between the pressure of the fluid within the hollow sphere and the tension of all the muscle fibers. The hollow sphere would be pushed apart at the cross-sectional plane if the sum of all the muscle fibers did not develop an opposing force of the same magnitude.

Therefore:
$$P\pi r^2 = nF, \tag{75}$$

where P is the inner pressure, πr^2 is the area of the sphere, F is the force of the individual muscle fibers and n is the number of fibers. Solving for F, this equation yields the force of the individual muscle fibers:

$$F = \frac{P\pi r^2}{n}. \tag{76}$$

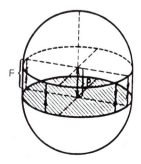

Fig 175 Heart is generally considered a hollow, spherical muscle. Inner pressure P attempts to separate two half-spheres from each other with the force $P\pi r^2$. This is counteracted by sum of force F of all muscle fibers around cross-sectional plane (according to M. Schneider).

It can be assumed from this equation that under identical conditions, the force of the individual muscle fibers increases with the square of the radius of the heart. This equation also shows that the development of force of the individual fibers is directly proportional to the inner pressure.

Thus, there are two ways to train the force of the myocardium: by increasing the pressure and by increasing the volume, i.e., the force of the heart can be trained by the product of the two variables, both of which represent a portion of myocardial work. Both factors can be of unequal magnitude with the same amount of myocardial work. A high blood pressure and thus also a high systolic intraventricular pressure are obtained during elevated work loads, as was seen on page 138. High continuous work loads cannot be maintained because they will rapidly produce fatigue of the working muscles. Intermittent exercise produces a marked increase in myocardial force because it is a form of isometric training for the myocardium. Of necessity, endurance does not need to be increased in this situation, i. e., endurance is dependent on the enzyme content of the mitochondria. Because the heart has little possibility to obtain energy from anaerobic glycolysis, it generally works aerobically. If myocardial cells go into an energy deficit, then frequent rhythm disturbances, marked pain and characteristic alterations of the ECG appear; these require that the work be interrupted because a myocardial infarction is threatening. In exactly the same way that was shown for muscle on page 279, maximal aerobic energy supply of the heart is dependent on enzyme content and oxygen supply. Myocardial oxygen supply is especially critical if the coronary arteries are narrowed due to the process of arteriosclerosis. Before beginning a training program for the heart, a good clinical evaluation of the heart must be made to determine whether it is healthy.

In the case of marathon training, i. e., training with constant, prolonged endurance exercise which is slightly above the threshold for prolonged work, the force of the heart is also trained, but obviously not to the same degree because enzymatic factors are more important. From the relationship shown in equation 76, it can be presumed that under identical conditions the force of individual muscle fibers increases with the square of the radius of the heart, insofar as the heart is assumed to be a simple sphere. Under these conditions, if the radius increases about 20%, then the corresponding increase in volume is about two-thirds of the initial volume. The increase in force of individual muscle fibers, under identical conditions, is about 44% of the initial force. This increased volume is closely correlated with the increase in blood volume. At the same peripheral blood pressure, the force of individual fibers will increase due to the enlarged volume.

In the case of a volume overload of the heart due to marathon training, peripheral blood pressure will be only slightly elevated. This means that the heart becomes smaller because the higher stroke volume is at the expense of the residual volume; in this case, the fiber lengths are at a minimum during the systolic maximum. Therefore, maximal tension development is less than when the heart contracts against pressure. In addition, there is the fact that during volume work, the isometric phase is shorter but the auxotonic phase is longer than during pressure work. With volume work, therefore, the heart is trained less for strength but more for endurance.

A heart that is well-trained for endurance is characterized primarily by its ability to oxidize large amounts of lactic acid during exercise, thus obtaining energy. The heart assures that the lactic acid formed during anaerobic work by skeletal muscles will be used during the work. As a result, the acid-base balance is not stressed as much. The endurance-trained heart has a much higher number and size of mitochondria compared to that of the untrained heart.

During a 43-day study on swimming rats, the effect of training on myocardial mitochondria was studied. Animals were killed at regular intervals after they swam 5–6 hours a day for 6 days a week. In the first days, an acute mitochondrial swelling was observed, with a disintegration of the mitochondrial cristae (p 10) and a thinning out of the matrix of most mitochondria. There was also the appearance of hypoxic damage to the myocardial cells. With more prolonged swim training, mitochondrial mass increased; this is expressed as the mitochondrial value determined by the ratio between mitochondrial proteins and myofilament proteins. This relation increased from 0.59 to 1.6 and then slowly returned to a value of 1.07.

Evidently, the myocardium is overloaded at the beginning of training. During the intervals between the 5–6 hours of swim training, the stress on the heart returned to a normal level. The recovery appeared to favor enlargement of the mitochondrial system. Similar relationships can also be seen for the training-related heart of the athlete, whose physiologic working capacity is increased and is not threatened by hypoxic damage.

The athletic heart is different from the heart pathologically overloaded by hypertension of different causes or by valvular stenoses and valvular insufficiencies and which must therefore continuously perform high levels of work. It is assumed that the average transverse diameter of a myocardial fiber is 16.2 μm. As a result, the total heart weight increases. If the fibers hypertrophy further, the diffusion pathway for oxygen becomes so great that the oxygen pressure is no longer sufficient to supply the mitochondria.

The critical heart weight at which this condition appears is stated to be 500 gm. As a result of this extreme fiber hypertrophy, a fiber insufficiency occurs. The athletic heart never reaches this condition because it does not surpass the critical weight of 500 gm. One therefore speaks

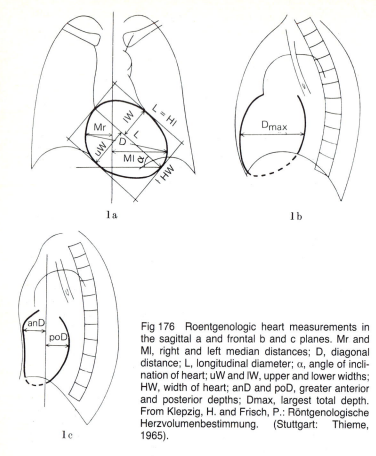

1 a   1 b

1 c

Fig 176   Roentgenologic heart measurements in the sagittal a and frontal b and c planes. Mr and Ml, right and left median distances; D, diagonal distance; L, longitudinal diameter; α, angle of inclination of heart; uW and lW, upper and lower widths; HW, width of heart; anD and poD, greater anterior and posterior depths; Dmax, largest total depth. From Klepzig, H. and Frisch, P.: Röntgenologische Herzvolumenbestimmung. (Stuttgart: Thieme, 1965).

of a proportional enlargement of all elements. The heart which is pathologically enlarged due to chronic pressure or volume overload and surpasses this value obviously forms new muscle fibers (hyperplasia) and new capillaries so that the supply of the heart is again guaranteed for a transitory phase. This occurs until these fibers once more reach the limit of their supply.

For the world of sports, it is important to know that an individual must be careful not to stress pathologically hypertrophied hearts with physical activity. A cardiologic investigation and medical advice are absolutely necessary in this case.

A frequent measure for determining heart size is the roentgenologically determined heart volume. As shown in Figure 176, the heart is x-

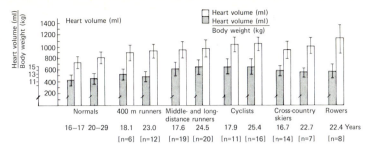

Fig 177   Absolute and relative heart size of young and adult normals, 400 m runners and endurance athletes. From Kindermann, W., Keul, J., Reindell, H.: Deutsche med. Wochenschrift 99 : 1372, 1974).

rayed in two planes, so that the measurements shown can be determined. From these measurements, heart volume can be estimated with the help of formulas (Klepzig and Frisch).

Values for the normal heart depend essentially on whether they are obtained with the subject lying or standing. Endurance training carried out during the growth period of ages 14–18 produces a rapid increase in heart size, such that the end result after a definite training period is the same as that obtained with adults. Figure 177 is an overview of the heart volumes of young and adult athletes in different disciplines. Next to each is the representation of relative heart volume (ml per kg body weight) in comparison to an untrained control group. The highest absolute heart volume is seen in rowers, demonstrating that this type of sport requires an especially large muscle mass. The relative heart volume is smaller than that seen in cyclists, because the increased weight due to muscle growth in cyclists is limited to the leg muscles.

Cardiac output at rest is not greatly altered by endurance training. The reduced heart rate is compensated by an increase in stroke volume. During maximal work loads, cardiac output can increase from about 15–20 l/minute with untrained persons to 25–35 l/minute with trained persons, depending on the level of training, type of training and body weight.

6.3.4.4 Effect of Endurance Training on the Increase in Blood Volume

During endurance training, blood volume is increased. There is a close relationship between the value of maximal O_2 intake and both blood volume and amount of hemoglobin. If prolonged endurance training is performed at sea level, blood volume will be increased 10–15% at the end of training, whereas hemoglobin concentration will be about 3%

less. An increased blood volume with approximately constant hemoglobin concentration means that there is an increase in total hemoglobin. With endurance training at altitude, both factors increase. In general, the increase in hemoglobin concentration is affected by training to only a small extent, the most important factor being altitude alone.

It is naturally attractive to speculate about the mechanism for increases in blood volume and hemoglobin due to training. Acute periods of strenuous physical exercise always produce a diminution in the amount of circulating blood and a measurable rise in the hematocrit value. This fact is understandable because such small molecular metabolites as lactic acid, pyruvic acid and others increase in muscle tissue and produce a displacement of water into the tissues due to osmotic pressure. As a result, blood loses water and its volume becomes less.

From investigations on immersion in thermoindifferent water (p 205), it is known that an apparent excess of intravascular volume causes an elimination of water and sodium and decreases plasma volume. This is associated with stretching of the atrial receptors due to the rise in central blood volume (Gauer-Henry reflex) and an inhibition of ADH and aldosterone secretion. In the case of endurance exercise with decreased blood volume, the opposite effect appears, namely, an increase in ADH and aldosterone secretion, with an increase in water and sodium retention, and a normalization of plasma volume with more prolonged exercise. After exercise, metabolites are rapidly broken down and water enters once more into the blood stream; this increased blood volume remains for 1–2 hours. The elevation in blood volume after exercise might lead to a slow adjustment of the receptor sensitivity in the atria so that blood volume is gradually increased. Hemoglobin concentration in the blood at rest clearly falls in the first 14 days of endurance training. This means that increased plasma volume is the first step; hemoglobin concentration is obviously only secondarily increased to approximately normal values.

The question must now be raised as to how the secondary normalization of hemoglobin concentration and the number of erythrocytes occurs. Both factors are regulated by an O_2 deficiency. It is probable that this O_2 deficiency does not affect bone marrow directly but via humoral factors (erythropoietin) which are initiated by local O_2 deficiency. Injection of plasma from persons living at altitude produces a constant formation of erythrocytes and hemoglobin in persons at sea level.

Erythropoiesis begins during residence at altitudes greater than 2,000 m, where the P_{O_2} in inspired air is less than 110 mm Hg. This means that the alveolar air then has a P_{O_2} of about 80 mm Hg. With a normal O_2 dissociation curve, there is a drop in saturation from 98 to 95% and the O_2 transport capacity is reduced from 21 to 20 vol%.

The same reduction in transport capacity of 100 ml blood at sea level can be obtained when blood volume increases about 7% due to higher water content with the same hemoglobin content, because hemoglobin concentration has now decreased from 16 to 15 gm/100 ml. If it is assumed that the sensor for erythropoiesis measures O_2 content of the blood, it is easy to imagine that erythrocyte production increases secondarily to the elevation in blood volume and that it continues only as long as necessary to again reach the desired value.

The composition of the sensor which measures O_2 content of the blood is not yet known. However, it could lie behind a tissue with constant blood flow and constant O_2 requirement and measure O_2 pressure there. One possibility for this would be the kidneys.

6.3.4.5 Effect of Endurance Training on Autonomic Functions

One of the most striking phenomena of an increased state of training is a lower resting heart rate. Whereas untrained persons have heart rates of about 70–80 beats per minute, endurance-trained, high-performance athletes often have heart rates of 35 beats per minute and less. As a result, vagotonia in individual cases can predominate to such an extent that there will be ECG changes similar to a bundle-branch block, i. e., there will be the sudden appearance of a secondary automatic center in the atrioventricular nodes. During exercise, the sinus nodes again assume the dominating role due to the sympathetic-induced frequency stimulation.

Arterial blood pressure at rest is approximately the same in trained and untrained persons. However, the regulatory characteristics of blood pressure are clearly diminished as a result of endurance training. Figure 67 (p 117) shows the mean values for blood pressure characteristics of 25 endurance-trained and 25 untrained subjects. The counter-regulation against a drop in blood pressure is distinctly reduced. The blood pressure servocontrol loop can be conceived as a type of PD (proportional-differential) servocontrol loop (Suppl, p 304). If the gain factor is calculated, its total range is reduced to about one-half the value for untrained persons. From a teleologic point of view, this finding is especially meaningful. It can be assumed that during exercise sympathetic tone from the muscle receptors is increased (p 133), so that blood pressure increases. As a result, the baroreceptors again reduce sympathetic tone. They are active in a "counter-coupling" manner. If their effect is diminished, so that the gain factor is reduced, the resulting effect is an elevated sympathetic tone and a wider adjustment range of the total integrated system. The adaptation of this system during exercise is gained at the expense of poor regulatory quality at rest. It should also be mentioned that trained individuals are much more sensitive to collapse than untrained individuals.

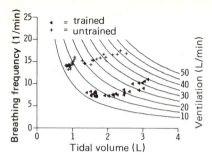

Fig 178
Relation between tidal volume and breathing frequency with endurance-trained (triangles) and untrained (pulses) persons, when ventilation at rest is stimulated by CO_2. Trained persons have lower breathing frequency and greater tidal volume at same ventilation.

The advantage of a high sympathetic tone is primarily in the greater ability to activate collateral vasoconstriction for distributing cardiac output between working muscles and inactive tissue.

The respiratory center is also affected by endurance training. Figure 178 shows that trained and untrained persons have the same ventilation at rest, but that it is obtained with different ratios of frequency and tidal volume when the respiratory center is stimulated at rest by CO_2. The same is true for the working ventilation of trained persons, who thereby require less oxygen for the same ventilation. As is clearly shown in Figure 179, the form of the CO_2 response curve of both groups is also different. As a rule, well-trained athletes have a reduced ventilatory response to the same alveolar CO_2 pressure. In addition, they also have a clear threshold, which is partly out of the range of physiologic CO_2 pressures.

6.3.4.6 Effect of Endurance Training on Acid-Base Balance and Mineral Content

The fact that the partially anaerobic working muscle constantly releases lactic acid and H^+ ions into venous blood means that the buffering capacity of the blood and thus the mineral balance are involved. When endurance-trained and untrained subjects perform the same work loads, the effect on the acid-base balance is related to how far the actual work load is above the individual's threshold for prolonged work. What is important is naturally not the work load itself, but the muscle metabolism necessary to perform it. The aerobic and anaerobic portions of this are also critical because the lactic acid equivalents released into the blood correspond to the degree of anaerobiosis. There are a number of mechanisms which can compensate for the H^+ ions that appear, partially via the buffering substances present in the blood and partially via the respiration and an increased CO_2 output (p 151). Respiratory work requires a significant fraction of

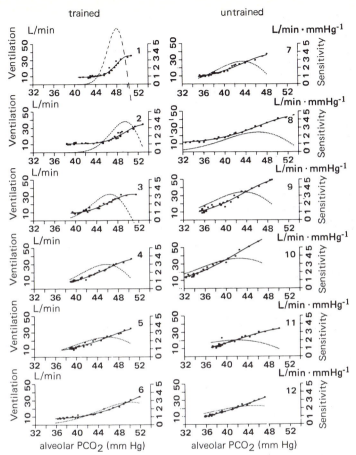

Fig 179   Relation between alveolar CO_2 pressure and ventilation for 6 endurance-trained and 6 untrained men at rest. As a rule, response is smaller in trained. Threshold of trained generally shows shift to the right. (Respiratory response curve, solid line; sensitivity, dashed line).

the total oxygen requirement during severe work loads. If ventilation must additionally compensate for a metabolic acidosis (p 154), then the circulatory system will be stressed even more by the transport of O_2 to the respiratory muscles. However, this fraction is not available for perfusion and as a result, the endurance capacity of the muscle is reduced. During the discussion on the effects of endurance training on

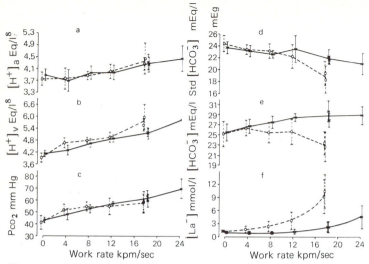

Fig 180   Pattern of several important chemical parameters in femoral venous and arterial blood of man during physical exercise. From Tibes, U., Hemmer, B., Schweigart, U., Böning, D., Fortescu, D.: Exercise acidosis as cause of electrolyte changes in femoral venous blood of trained and untrained men. Pflügers Arch. 341 : 145–158, 1974).

the heart, it was determined that the trained heart was capable of utilizing large amounts of lactic acid; the same is true for inactive muscles. The discussion now centers on several mechanisms which support physical performance capacity during endurance training.

First of all, we will consider how several relevant parameters in femoral venous blood (i. e., in the blood coming from the working muscles) and in arterial blood are affected during bicycle ergometer exercise (Fig 180). The values were obtained from 6 highly-trained athletes in endurance sports and from 6 completely untrained subjects. Work loads were progressively increased up to 18 kpm/second for untrained subjects and up to 24 kpm/second for the trained subjects; the duration at each level was 10–15 minutes. Blood samples were taken at the end of each work load.

The P_{CO_2} of femoral venous blood (Fig 180, C) increased more or less linearly with work load in the trained subjects, while the increase above 4–8 kpm/second in the untrained subjects was small. Lactate (Fig 180, F) in femoral venous blood was less in trained subjects even at rest and increased less markedly when compared at the same workload. The difference at the higher work loads was especially

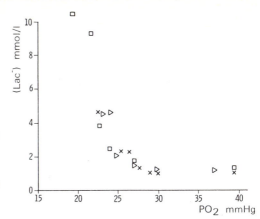

Fig 181
Relationship between
PO$_2$ and lactate con-
centration in femoral
venous blood at
different work loads in
man (according to
Tibes).

noticeable. Considering the standard bicarbonate (Fig 180, D) and the actual bicarbonate concentration (Fig 180, E), trained subjects at higher work loads reacted more with a respiratory acidosis, whereas untrained subjects reacted more with a metabolic acidosis. The hydrogen ion concentration in femoral venous blood (Fig 180, B) increased less in the trained subjects than in the untrained subjects in relation to the same work loads. Figure 180, A shows that during the flow of blood through the lungs, a portion of the H$^+$ ions was evidently intercepted, for [H$^+$] in arterial blood was less in both groups.

The reason for the different lactate production is multifactorial. First, it is certain that the modified enzyme activity described on page 282 is the determining factor in trained persons. Furthermore, there is a close relationship to oxygen pressure (Fig 181) in the femoral venous blood; this represents a local integral of the individual capillary O$_2$ pressures. The relationship between lactate concentration and P$_{O_2}$ is obviously the same for differing levels of training. Perfusion is higher in the regions of low O$_2$ pressure, so that lactate formed will dissolve in more blood.

According to the Fick law, under identical conditions, the pressure difference between blood and the mitochondria is decisive for oxygen supply to the mitochondria. Therefore, a higher P$_{O_2}$ in the capillaries is important during high levels of metabolism. The endurance-trained individual has an O$_2$ dissociation curve which is shifted to the right, resulting in a higher O$_2$ pressure in the tissues at the same saturation. Within the decisive range of lower saturation, the influence of the volatile acid H$_2$CO$_3$ on the displacement to the right of the O$_2$ dissociation curve (Bohr effect, p 94) is stronger than that seen in fixed acids when both are compared at the same [H$^+$]. The respiratory acidosis appearing in the capillaries of trained subjects is more effec-

Table 23 Mean Values of Cations and Orthophosphate Alterations in Femoral Venous Blood. For Correction of (Na^+), see Table 24. From Tibes, U., Hemmer, B., Schweigart, U., Böning, D., Fotescu, D.: Pflügers Arch. 347: 145–158, 1974.

Work load kpm/sec		0	4	8	12	18	18	24
$[Mg^{2+}]$ mEq/l	UT	1.51±0.09	1.54±0.09	1.56±0.08	1.58±0.08	1.61±0.09	1.58±0.07	
	TR	1.43±0.08	1.42±0.07	1.41±0.09	1.44±0.07	1.44±0.07	1.45±0.08	1.47±0.08
	ST	1.48±0.07	1.50±0.07	1.57±0.13	1.53±0.10	1.54±0.09	1.56±0.07	
$[Ca^{2+}]$ mEq/l	UT	4.95±0.29	5.02±0.35	5.16±0.39	5.16±0.36	5.34±0.44	5.50±0.39	
	TR	5.13±0.33	5.16±0.16	5.40±0.39	5.31±0.26	5.25±0.13	5.74±0.13	5.52±0.15
	ST	5.68±0.08	5.76±0.21	5.61±0.11	5.73±0.16	5.91±0.08	5.84±0.16	
$[Na^+]$ mEq/l	UT	139.50±1.60	140.40±1.10	140.20±0.90	140.60±1.50	141.90±1.40	142.00±1.20	
	TR	139.90±1.30	139.80±1.20	140.40±1.10	140.40±1.30	141.00±1.00	141.00±1.20	141.90±1.30
	ST	138.90±1.00	139.50±0.80	138.10±1.60	139.60±2.30	142.40±2.00	140.90±1.20	
$[K^+]$ mEq/l	UT	3.74±0.13	4.02±0.07	4.23±0.10	4.29±0.08	4.59±0.18	4.60±0.13	
	TR	3.81±0.10	4.00±0.13	4.06±0.11	4.13±0.14	4.27±0.17	4.27±0.17	4.49±0.29
	ST	3.91±0.09	4.03±0.05	4.09±0.11	4.18±0.10	4.32±0.12	4.36±0.06	
[Pi] mmol/l	UT	0.84±0.21	0.92±0.22	1.05±0.20	1.19±0.19	1.28±0.21	1.44±0.27	
	TR	1.00±0.17	1.05±0.24	1.18±0.25	1.26±0.25	1.34±0.21	1.39±0.25	1.55±0.25
	ST	0.79±0.22	.99±0.30	1.01±0.33	0.95±0.40	1.03±0.40	1.07±0.45	

UT: untrained; ST: semitrained

Table 24 Correction Factors Needed to Determine Work-Related Alterations in Various Plasma Components in Relation to Plasma Volume. (Na$^+$) in Table 23 was Determined by Multiplication with the CF$_{Hct}$. From Tibes, U., Hemmer, B., Schweigart, U., Böning, D., Fotescu, D.: Pflügers Arch 347: 145–158, 1974.

Work load kpm/sec		4	8	12	18	18	24
CF$_{Hct}$	UT	0.985	0.981	0.974	0.947	0.936	0.952
	TR	0.991	0.985	0.977	0.967	0.957	
	ST	0.990	0.983	0.969	0.954	0.955	
[Na$^+$] · CF$_{Hct}$ mEq/l	UT	138.3±2.6	137.5±1.5	136.9±1.7	134.5±2.9	133.2±2.3	
	TR	138.6±1.7	138.2±3.2	137.5±2.9	136.4±2.5	134.9±2.2	135.1±1.6
	ST	138.1±0.6	135.8±4.0	135.3±1.7	136.3±3.0	134.8±0.4	

UT: untrained; TR: trained; ST: semitrained

tive for O_2 supply to the muscles than the metabolic acidosis of untrained subjects. The total displacement to the right of the O_2 dissociation curve in trained subjects cannot be explained completely either by the Bohr effect or by the changes in 2,3-DPG content (p 95) and additional unknown factors must be assumed. A shift to the right supports the performance of endurance exercise.

There are also marked differences between trained and untrained subjects in electrolyte concentrations in femoral venous blood during bicycle ergometer exercise. It must be remembered that the definition of concentration is always the amount of a substance divided by the volume of the solvent. If the same amount of substance is in a lesser amount of solvent, concentration rises without any increase in the amount of the substance.

A number of small molecular substances (lactate, glucose degradation products) appear in the working muscle cell; the cell membrane represents a permeability barrier for these substances. Due to osmotic pressure differences, water leaves the blood stream and enters the cell, so that blood and all of its components become more markedly concentrated. However, erythrocytes do not leave the blood stream during exercise. For this reason, the increase in hematocrit can be used as a measure of hemoconcentration.

Table 23 shows the concentration of several important cations and of inorganic phosphate in femoral venous blood at different work loads from the above-mentioned series of investigations. The correction factors explained previously are shown in Table 24. After considering the correction factors, it is possible to calculate that Mg^{++} and Ca^{++} maintain essentially the same concentration, i. e., that magnesium and calcium are not released in increased amounts from muscle. The $[K^+]$ and $[Na^+]$ react differently in the different groups. Sodium ion and potassium ion concentrations in the blood are related to hydrogen ion concentration, because it is possible that the greater the hydrogen ion concentration, the more the sodium-potassium pump will be inhibited. Since the $[H^+]$ in untrained subjects rises more at the same work loads, $[K^+]$ exhibits the analogous pattern. From this, it can be assumed that additional potassium leaves the muscle cell and increased amounts of sodium enter the muscle cell when the hydrogen ion concentration is greater. The relationship between $[K^+]$ and the control of sympathetic tone has already been mentioned (p 138).

7 Basic Concepts of Biologic Regulation (Biologic Cybernetics)

Even though the conceptual model on which biologic regulation theory is based was first used in physiology in the previous century, its formal consideration is only about 50 years old and has been more completely developed within the last 20 years. The term cybernetics is derived from the Greek work "kybernetes" (helmsman) and was coined by the American scientist Norbert Wiener. Biologic cybernetics is comprised not only of the area of biologic control but also of the treatment of information within the organism. This latter aspect cannot be discussed in detail here.

The application of regulation theory in the organism has several essential tasks. First, there is a scientific task because technologically developed research methods and mathematical evaluation procedures of the regulation theory can be adopted to make the characteristics of a biologic system accessible to quantitative investigation. In addition, there is a didactic task, because complicated relationships can be represented in system-independent, abstract block diagrams, such that the characteristics of a block must be known only in their input-output relationships. The person who enters into this system first of all gains a general overview without having to lose himself in details. In addition, the biologic servocontrol loop has a heuristic value in that questions can be raised about parts of the system, without which the system cannot function. In another sense, the biologic servocontrol loop is a simplified abstract reflection of biologic reality.

7.1 Construction of a Servocontrol Loop

The construction of a servocontrol loop is objectively represented in Figure 182, using the conventional nomenclature of control technology. The task of the apparatus is to maintain the water level (controlled variable X) in the container independent of a variable amount of inflow (disturbance load variable Z) at a constant and prescribed level. A necessary condition is that the level of water is continually being measured. This measurement is effected by a sensor. Information about the position of the sensor (i. e., the actual value of the manipulated variable) is conveyed to the control center, as symbolized by the attendant. The control center compares the actual value with the desired value or set point, which is prescribed by the reference input variable W. Using this plan, the reference input can be held

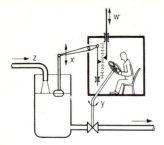

Fig 182   Manual control of fluid level. From Oppelt, W.: Kleines Handbuch technischer Regelvorgänge (Weinheim: Verlag Chemie, 1972).

constant (homeostat) or can be varied according to a prescribed function (servomechanism). If the actual value corresponds to the set point, then there will be no reaction. As soon as there is a difference between the actual and the desired levels (control deviation), then the correction device of the regulator will be activated; this modifies the manipulated variable Y and thus the outflow from the container. As a result, negative feedback is introduced. If the actual value is increased, then the regulator increases the outflow even more. The portion of the servocontrol loop described above is called the regulator. It consists of a sensor, an amplifier (control center) and a correction device. The portion on which the regulator acts is called the controlled system. It describes the effect of the correction device on the input of the regulator.

Figure 183 shows a controller which functions automatically without human help. It is a characteristic of a servocontrol loop that information flows in only one direction. The further and final abstraction is represented by the block diagram in Figure 184, in which the regulator (R) and controlled system (C) are represented only in the form of boxes. The block with the symbol R represents the regulator, whose input represents the sum of the sensor information (X) and the reference input (W), by which X and W can assume negative as well as positive values. The regulator manipulates the sum of the input according to a definite mathematical function, as we shall see in the discussion of the dynamics of servocontrol loops. What is important is that the output information is made negative, i. e., the correction device is always activated in the opposite direction to the sign of the input. Thus, there is a negative feedback, without which the servocontrol loop could not function. A positive feedback would produce a vicious circle, i. e., a continually greater deviation from the set point.

By means of formal manipulation, the influence of the disturbance input Z is added to the output information of the servocontrol loop; the disturbance input Z will then be regulated. The controlled system follows another mathematical function, depending on its physical

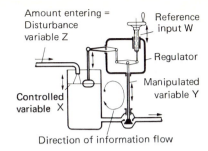

Fig 183
Automatic control of fluid level.
From Oppelt, W.: Kleines Handbuch technischer Regelvorgänge (Weinheim: Verlag Chemie, 1972).

composition. The output information of the regulator is coupled to its input via this function.

7.2 Characteristics of a Technical Regulator

Technical regulators are divided into three main groups (Fig 185). Each of these regulator groups has advantages and disadvantages. Their use is governed essentially by the regulator problem, that is, by the physical characteristics of the controlled system. In biology as well, the same measurement principles are realized by different biologic servocontrol loops. The main groups are characterized by the mathematical relationship between the input of the servocontrol loop,

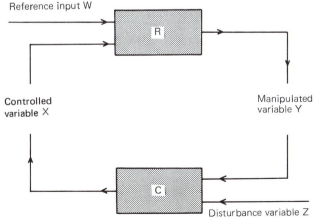

Fig 184　Block diagram of an automatic control system.

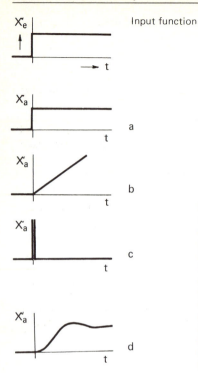

Input function

Fig 185   Transfer responses of servomechanism components. a, P elements which give output signal proportional to the input signal; b, I elements which integrate time course of input signal; c, D elements which differentiate course of input variable over time, thus forming rate of change; d, effects of delay allow description of input pattern which is delayed at output. From Oppelt, W.: Kleines Handbuch technischer Regelvorgänge (Weinheim: Verlag Chemie, 1972).

the manipulated variable X and the effector Y, which activates the correction device.

7.2.1 P Regulator

The term P regulator stands for the proportional regulator. In this type, the change in the effector is proportional to the change in the manipulated variable, i. e., there is a proportional relationship between the measured variable X and the effector Y. This type of regulator built into a servocontrol loop causes a constant error which depends on the proportionality factor between X and Y.

7.2.2 PD Regulator

The term PD regulator stands for proportional-differential regulator. This servocontrol loop acts in the steady state as a P regulator. The state of adjustment represents a system error for the effector and the speed of the system error alteration is decisive.

7.2.3 I Regulator

The term I regulator stands for integral regulator. The correction device (usually a servomotor) adjusts itself with a speed proportional to the system error. The value of the effector corresponds to the time integral of the system error. This type of regulator is more complicated to construct and exhibits no system error. When there is inadequate adjustment, it has a tendency to oscillate.

In addition, there are mixed forms, such as PI and PID regulators.

7.2.4 Transfer Function of the Open Servocontrol Loop

A servocontrol loop can be interrupted at an arbitrary site. In this way, the same dimensions can be obtained at the input and output, so that the static (steady-state relationship) and the dynamic response of the open loop can be evaluated. With most linear systems, there is the possibility of predicting the response of the closed loop from the response of the open servocontrol loop. First of all, consider a simple open servocontrol loop with a P regulator. As shown in Figure 185, the ideal response ($X_{a(t)}$) is that which corresponds to each time point, t:

$$X_{a(t)} = V_O \cdot X_{e(t)}. \tag{77}$$

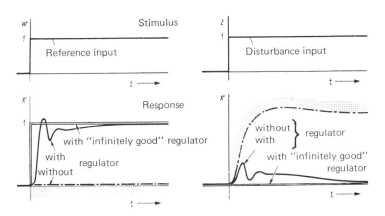

Fig 186   Typical responses of control system to stepwise input of reference or disturbance variable. Notice limiting states "without regulator" and "with infinitely good regulator". From Oppelt, W.: Kleines Handbuch technischer Regelvorgänge (Weinheim: Verlag Chemie, 1972).

This is tested with a stepwise function of X_e (input) and the variable X_a (output) is recorded. In actuality, the regulator is composed of physical or biologic components which react more or less slowly. In addition, there is frequently a larger or smaller dead time. This means that the transmission of the information requires a certain time (e. g., nerve conduction velocity, transport of information via the blood stream), so that equation 77 actually assumes another form, which we will only graphically represent for the sake of simplified understanding. For $t \rightarrow \infty$, the servocontrol loop function surpasses the ideal function of equation 77.

From Figure 185c, the meaning of the derivative (D) elements that compensate for a part of the slow response reaction can be perceived. The form of the transient response gives information about the dynamic responses. During the steady-state condition ($t \rightarrow \infty$), the open-loop gain V_0 can be determined by:

$$V_O = \frac{X_a}{X_e}. \tag{78}$$

From this, the control factor can be obtained from the following relationship:

$$R = \frac{1}{(1 + V_O)}. \tag{79}$$

From this, the precision of the closed servocontrol loop can be determined because the disturbance input Z (which it is supposed to be compensated) is regulated only with the product RZ. From equation 79, it can be seen that the static regulatory factor can take on values between 0 ($V_0 = \infty$) and 1 ($V_0 = 0$). When $R = 1$, the full disturbance load remains active; when $R = 0$, the disturbance load is completely regulated. Practically speaking, however, there is an instability, as we shall see.

7.2.5 Polar Plot of the Frequency Response of the Open Servocontrol Loop

The polar plot of the frequency response can be obtained if the open servocontrol loop is stimulated at input frequencies of constant amplitude, instead of with a stepwise stimulation. With a linear system, the output X_a is also sinusoidal and only in the case of the ideal equation (77) would it be a function with a phase angle of 0 and a constant amplitude. However, dead time and delay time of a simple P system produce a characteristic phase angle and a characteristic amplitude for each frequency. These are appropriately recorded in a complex plane, as represented in Figure 187. In this way, the phase angle, starting from the real positive axis, is recorded in the clockwise direction with an arrow, the length of which corresponds to the amplitude. The polar plot is obtained in this way.

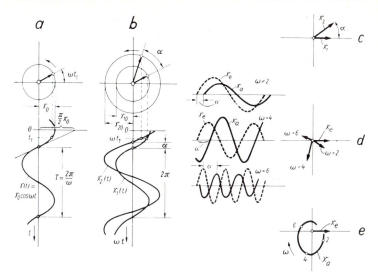

Fig 187 Origin of polar frequency response locus from recording of harmonic oscillations. From Oppelt, W.: Kleines Handbuch technischer Regelvorgänge (Weinheim: Verlag Chemie, 1972).

From the form of the polar plot, it is apparent which type of regulator is involved and whether the system has any dead time; in addition, it is evident whether the closed servocontrol loop is stable.

7.2.6 Nyquist Stability Criteria

Figure 188 shows a polar plot where the reference input (F_o) of a PD servocontrol loop was modified in a sinusoidal manner. The reaction of the servocontrol loop was stable. The stability criterion of Nyquist is: if

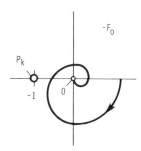

Fig 188 The Nyguist stability criterion (see text for more details).

one moves on the polar plot in a clockwise direction, then the critical point (P_c) must always lie on the left side. The critical point is that point which lies on the real negative axis at a point symmetric to the input amplitude of the real positive axis. The stability criterion is evident because negative feedback is functionally converted in a positive direction by a phase shift of 180 degrees. A positive feedback is the basis of each oscillating circuit which impinges on itself.

7.3 Biologic Servocontrol Loops

Technical and biologic servocontrol loops basically work according to the same principles. The decisive difference from the viewpoint of biology is that the engineer constructs the regulator primarily according to the requirements of the task and essentially knows its characteristics beforehand, while the biologist must analyze servocontrol loops which are present in order to understand their characteristics. Control theory principally deals with linear systems, whereas the biologist predominantly contends with nonlinear systems.

8 Methods of Energy Metabolism Measurement

As has already been mentioned, direct calorimetry has no practical role. Therefore, we can limit our discussion to the most widely used methods of indirect calorimetry. The theoretical relationship between O_2 intake and energy turnover was discussed on page 60.

8.1 Open Systems

An open system is characterized by arrangements in which the subject breathes in air from the environment via a valve and expires via another valve into a collection system, so that both the amount and composition of the expired air can be determined. The classic collection system is represented by the Douglas bag (Fig 189), which is made of either rubberized canvas or plastic material.

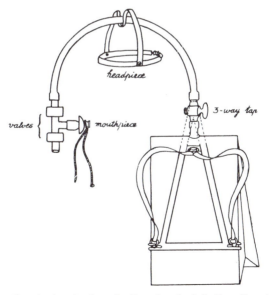

Fig 189   Douglas bag for the collection of expired air. From Douglas, C. G.: J. Physiol., London 42 : 17, 1911).

The most commonly used sizes have a collection volume of 100–250 l. The stopcock found directly on the bag is usually constructed as a 3-way stopcock, so that the bag is closed in the period prior to collection although the subject is already breathing out through the connecting tube. At the beginning of the measurement, time is determined and the 3-way valve is turned so that the expired air is now collected in the bag. At the end of the study period, the time is determined once more and the bag is closed.

Immediately after the study, the content of the Douglas bag is mixed well and pumped through a gas meter with a temperature indicator. After approximately half the gas has gone through the gas meter, a sample will be taken through the tube on the side for chemical gas analysis. The volume measured by the gas meter divided by the time of collection gives the timed volume, which must then be reduced to normal conditions (0 C, dry, 760 mm Hg) to obtain the "reduced ventilation." To do this, the measured gas volume must be multiplied by a correction factor (f). It is determined according to the following relationship:

$$f \text{ (wet)} = \frac{P_B - P_{H_2O}}{760 \, (1 + \alpha t)}. \tag{80}$$

where P_B and t = barometric pressure and temperature in mm Hg and degrees C, respectively, within the measured air volume, P_{H_2O} = water vapor pressure at the temperature t, and α = the cubic expansion coefficient of air ($\alpha = 0.00367$).

The correction factors can also be found in tables (e. g., Documenta Geigy).

Chemical gas analysis yields the per cent composition of the expired air, which in this case is assumed to be 2.86 vol% CO_2 and 17.64 vol% O_2. This analysis is usually done now with the Scholander apparatus. To determine O_2 utilization, the following calculation must be applied. At an R less than 1, less CO_2 is breathed out than O_2 is breathed in and the volume of expired air is less than that of the inspired air. This "shrinkage" of respiratory air is expressed in the increased N_2 content because the amount of N_2 remains unchanged and is eliminated. By subtracting the sum of per cent O_2 and per cent CO_2 from 100, the per cent N_2 content of expired air can be found. One can now calculate how high the per cent O_2 content would be increased due to the shrinkage of the respiratory air, if no O_2 had been retained in the body. This is possible because there is a fixed relationship of O_2 to the sum of N_2 + inert gases equal to 0.265; the N_2 content of the expired air is thus multiplied by 0.265. To simplify the calculation process, values for (100 −[per cent O_2 + per cent CO_2]) × 0.265 can be found in Table 25.

An example should make the calculation process clear once again. The measured values are:

Table 25 Calculation of Reduction in Volume of Inspired Air (see Text)

$[O_2\% + CO_2\%]$	$(100-[O_2\% + CO_2\%]) \cdot 0.265$	$[O_2\% + CO_2\%]$	$(100-[O_2\% + CO_2\%]) \cdot 0.265$	$[O_2\% + CO_2\%]$	$(100-[O_2\% + CO_2\%]) \cdot 0.265$
20.00	21.2000	20.30	21.1205	20.60	21.0410
20.01	21.1973	20.31	21.1178	20.61	21.0383
.02	.1947	.32	.1152	.62	.0357
.03	.1920	.33	.1125	.63	.0330
.04	.1894	.34	.1099	.64	.0304
.05	.1867	.35	.1072	.65	.0277
20.06	21.1841	20.36	21.1046	20.66	21.0251
.07	.1814	.37	.1019	.67	.0224
.08	.1788	.38	.0993	.68	.0198
.09	.1761	.39	.0966	.69	.0171
.10	.1735	.40	.0940	.70	.0145
20.11	21.1708	20.41	21.0913	20.71	21.0118
.12	.1682	.42	.0887	.72	.0092
.13	.1655	.43	.0860	.73	.0065
.14	.1629	.44	.0834	.74	.0039
.15	.1602	.45	.0807	.75	.0012
20.16	21.1576	20.46	21.0781	20.76	20.9986
.17	.1549	.47	.0754	.77	.9959
.18	.1523	.48	.0728	.78	.9933
.19	.1496	.49	.0701	.79	.9906
.20	.1470	.50	.0675	.80	.9880
20.21	21.1443	20.51	21.0648	20.81	20.9853
.22	.1417	.52	.0622	.82	.9827
.23	.1390	.53	.0595	.83	.9800
.24	.1364	.54	.0569	.84	.9774
.25	.1337	.55	.0542	.85	.9747
20.26	21.1311	20.56	21.0516	20.86	20.9721
.27	.1284	.57	.0489	.87	.9694
.28	.1258	.58	.0463	.88	.9668
.29	.1231	.59	.0436	.89	.9641
.30	.1205	.60	.0410	.90	.9615
		20.91	20.9588		
		.92	.9562		
		.93	.9535		
		.94	.9509		
		.95	.9482		
		20.96	20.9456		
		.97	.9429		
		.98	.9403		
		.99	.9376		
		21.00	.9350		

1. Reduced ventilation per minute   14 l/minute
2. Concentration of O_2 in the expired air   17.64%
3. Concentration of CO_2 in the expired air   2.86%

The sum of 2 and 3 is 20.50.

According to this, the concentration of nitrogen + inert gases in expired air is $100 - 20.5 = 79.5\%$.

The apparent concentration of O_2 in inspired air is therefore $79.5 \times 0.265 = 21.07\%$.

Required O_2 per cent: $21.07 - 17.64 = 3.43$.

$$R = \frac{2.86}{3.43} = 0.83.$$

$$\text{Required oxygen (ml/minute)} = \frac{3.43 \times 14000}{100} = 480 \text{ ml/minute}$$

$$CO_2 \text{ given off (ml/minute)} = \frac{2.86 \times 14000}{100} = 400 \text{ ml/minute}$$

Metabolic turnover = 480 ml O_2/minute $\times$ 4.838 kcal/l O_2 (caloric equivalent) = 2.32 kcal/minute (compare with Table 2).

An improvement on the method of the open system is represented by the respiratory gas meter of the Max Planck Institute of Work Physiology. As in the case of the Douglas bag, the subject breathes out via a respiratory valve directly into the gas meter. A small pump activated by the gas meter itself extracts an aliquot sample (selectively 0.3% or 0.6%) of the expired volume in an anesthesiology bag for later analysis (Fig 190). This aliquot has the following significance. Because the sample volume drawn out by the small pump is proportional to the expiratory velocity, the per cent fraction of dead space, mixed air and alveolar air has the same concentration in the anesthesiology bag as in the expired air. Gas volumes are measured directly via the gas meter, which also contains a thermometer, and the per cent composition of expired air in the anesthesiology bag is chemically determined. The calculation of energy turnover is the same as in the Douglas bag procedure.

The open system is appropriately used where energy metabolism must be determined outside of the laboratory because both the Douglas bag and the respiratory gas meter are portable. The respiration gas meter is preferable to the Douglas bag because it requires less space. In addition, the amount of gas can be read directly at the working place. With both methods, however, one should be aware of the fact that the gas concentration in rubberized canvas bags, as well as in anesthesiology bags, changes with time because CO_2 diffuses through rubber. In the case of plastic bags, CO_2 diffusion is markedly dependent on the

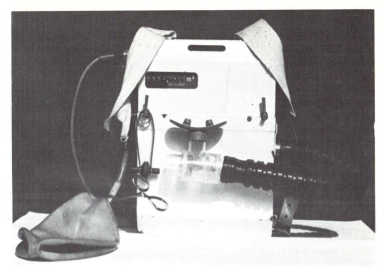

Fig 190   Respiratory gas meter from the Max Planck Institute for Work Physiology in Dortmund (Müller, E. A. and Franz, H.).

concentration of the plasticizer used. It is recommended in all cases that the gases should be analyzed as soon as possible or, if this is not feasible, the gas samples should be transferred to evacuated glass ampules until such time as the analysis can be carried out.

False energy metabolism values can be obtained using the open system if changes in the composition of inspired air go unnoticed. The oxygen content in room air normally corresponds to that of exterior air. However, if energy metabolism measurements are carried out in an oxygen deficiency or near combustion ovens, then the inspired air must also be simultaneously measured to avoid errors. During investigations on forestry workers in the woods, it should be remembered that oxygen content can be higher than normal due to intense sunshine, especially when there is little air movement.

8.2 Metabolic Measurements with the Closed System

Metabolic measurements obtained with the closed system will be illustrated with the help of the Krogh spirometer (Fig 191). The spirometer lever encloses an air space with a surrounding water jacket. The volume change in the air space due to respiratory movement is

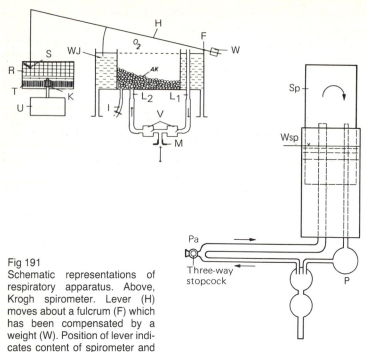

Fig 191
Schematic representations of respiratory apparatus. Above, Krogh spirometer. Lever (H) moves about a fulcrum (F) which has been compensated by a weight (W). Position of lever indicates content of spirometer and is continuously recorded on a kymograph. Right, principle of Knipping spirograph; WJ, water jacket; SL, soda lime to absorb CO_2; V, valve; I, inlet; P, pump; Sp, spirograph bell.

recorded by the movement of the lever around the fulcrum (F). Via the respiratory valve, air flow is linked in such a way that gas flow from the spirometer is carried into the lungs of the subject and then once more into the spirometer via the carbon dioxide-absorbing mass of soda lime. The volume reduction of oxygen in the time period is recorded and reduced to normal values. Because CO_2 cannot be determined using this arrangement, an average R of 0.85 is assumed in order to calculate energy metabolism.

Apparatus measuring metabolism in a closed system is normally used in the clinical situation due to its simple operation. This procedure has been perfected, especially by Fleisch and Knipping. A diagram of the principles of the Knipping apparatus is shown in Figure 191. It is obvious that metabolic measurements with the closed system can be carried out only in laboratories with an apparatus that is permanently installed. The disadvantage of the closed system is especially evident when each change in respiratory midpoint is registered as an apparent change in oxygen intake. Some of the closed systems have a disturbing effect because the gas composition of the inspired air is not constant; this can be eliminated by using expensive equipment with an oxygen stabilizer.

8.3 Computer-Controlled Spiroergometry using the Method of Breath-by-Breath Analysis

The development of measuring techniques today makes it possible to revise the classic procedures, such that the results can be obtained faster, with more precision and also with more significant parameters being measured simultaneously. Due to the marked rise in cost of health care, the scientific viewpoint has an increasingly important role for medical check-ups or routine investigations. In this regard, personnel costs which occur due to manual evaluation and representation of results are particularly important.

In the area of training research, it is necessary to be able to study as many athletes as possible and at the same time to keep the work involved for individual investigations the lowest possible for the subjects and for the personnel carrying out the investigation. One means of doing this is offered by on-line, electronic data processing with the aid of a process computer, where the course of the experiment and the evaluation can be automatically controlled by the computer. Such automatization also means the utilization of comparable and standardized methods, so that a better comparability of results among different investigative centers can be maintained. Another aspect is that data documentation can be performed directly on digital tape or magnetic discs, without the intermediate use of punched cards and the care required for them. Data from later investigations can be compared to the earlier results with little expense. In addition, they can be more easily recovered for statistical and epidemiologic purposes.

8.3.1 Theoretical Foundations of the Procedure

If expired air is divided into equal volume fractions, then each contains a different gas concentration than the previous one. The first fraction practically corresponds to inspired air because it filled only dead space and did not participate in the gas exchange. With each additional fraction, the gas concentration approaches that of alveolar air. If the expired volume (V_E) is divided into n equal subvolumes (ΔV_i; $i = 1, 2 \ldots n$):

$$V_E = \Delta V_1 + \Delta V_2 + \ldots \Delta V_n, \tag{81}$$

then the CO_2 volume given off during this breath is equal to the momentary concentration difference between inspired and expired air Ci:

$$V_{CO_2} = \Delta V_1 \cdot \frac{C_1}{100} + \Delta V_2 \cdot \frac{C_2}{100} + \ldots \Delta V_n \cdot \frac{C_n}{100}. \qquad (82)$$

$\frac{C}{100}$ is called the fraction F.

To obtain time segments from the volume fractions which are equal and technically simpler to control, each section in equation 82 is enlarged with Δt and is thus:

$$V_{CO_2} = \frac{\Delta V_1}{\Delta t} \cdot F_1 \cdot \Delta t + \frac{\Delta V_2}{\Delta t} \cdot F_2 \cdot \Delta t + \ldots \frac{\Delta V_n}{\Delta t} \cdot F_n \cdot \Delta t \qquad (83)$$

or with $\Delta t \rightarrow 0$:

$$V_{CO_2} = \frac{dV_1}{dt} \cdot F_1 \, dt + \frac{dV_2}{dt} \cdot F_2 \, dt \ldots + \frac{dV_n}{dt} \cdot F_n \, dt \qquad (84)$$

or using the summation sign:

$$V_{CO_2} = \sum_{i=1}^{n} (\frac{dV_i}{dt} \cdot F_i \, dt). \qquad (85)$$

The respiratory flow intensity $\frac{dV_i}{dt}$ is appropriately recorded with the Fleisch pneumotachograph.

The measuring principle is essentially as follows. Respiratory air flow is laminarized via a system of lamina, obeying the Hagen-Poiseuille Law (p 101). Because viscosity depends a great deal on temperature, the lamina are heated to a constant value. In this way, the problem of possible fluid condensation is avoided, thereby changing the defined respiratory resistance. The pressure difference before and after the resistance is measured via an electric transducer.

Using this measurement principle, the volume V is obtained by integration of flow intensity $\frac{dV_i}{dt} = \dot{V}$ over time. This is now produced either in the analog form with an integrator or the digital form with a computer.

8.3.2 Practical Performance of the Measurement with a Process Computer, a Mass Spectrometer and a Pneumotachograph

We use a respiratory mass spectrometer where three channels continually measure concentrations of O_2, CO_2 and N_2. Other forms of low-inertia concentration-measuring equipment can also be used. The intensity of respiratory flow is recorded with a pneumotachograph. We also have a digital process computer available. This type of computer works with real numbers, which the machine, similar to an electronic calculator, can then recalculate, produce, draw and graph.

Because both the mass spectrometer and the pneumotachograph produce continuous, analog (proportional to the measuring reaction) voltage, the first step is to transform this into real values (digitalize). This is previously programmed via an analog-digital converter. The analog signal of the pneumotachograph is proportional to the respiratory flow intensity. The analog signal from each channel of the mass spectrometer is proportional to the corresponding partial gas pressure. From equation 84, it can be seen that a definite rhythm is necessary which corresponds to dt. To obtain the necessary precision of the summation, a programmable quartz clock is used which causes the converter at each equal time interval dt to digitalize the input signal. In the previous case, four channels are needed. The duration for each conversion is 16 μsec per channel. The duration of the interval dt was selected to be 10 msec. The computer not only converts the numerical values but must already have completed all calculations before the next interval because the results are to be displayed at the end of each breath.

8.3.3 Calibration of the Individual Measurements

To calibrate gas concentration, test gases for N_2, O_2 and CO_2 with a degree of purity of 99.99% are used and mixed with a precise gas mixing pump, producing 4% and 8% CO_2 in N_2 and 14% O_2 in N_2. The dry mixtures are drawn one after another into the mass spectrometer and recorded on the computer. Because the gases are admitted via a magnetic valve, which is controlled by the computer, the user does not have to be concerned about sufficient equilibration time. It has already been incorporated within the program.

Because the indication of each channel is linear with the partial pressure, the calibration constants a and b of each can be calculated from an equation which has the form:

$$y = ax + b \tag{86}$$

(x = indicated value, y = calibrated value).

With this, the corresponding partial pressure or, with the help of the actual barometric pressure, the concentration of the reference gases can be obtained. The different calibration constants a and b for each gas measured by the calibration program are determined and stored. To be independent of the water vapor pressure in the expired air (this is always a well-known problem with the mass spectrometer because the inspiratory and expiratory values are unequal), the following algorithm is used. The sum of the partial pressures of the dry gases N_2,

O_2, CO_2 and argon is practically the same as the total pressure or related to the concentration, as a calculated value of 100%. If argon is not considered, then the sum is 99.94%. If there are no additional foreign gases or water vapor, then this sum remains constant. The effect of water vapor pressure is eliminated because each partial gas pressure is always multiplied by the following factor (K):

$$K = \frac{0.994}{F_{CO_2} + F_{N_2} + F_{O_2}}, \tag{87}$$

where F is the fractional portion of gas volume. In this way, gas pressures and concentrations are continuously shown under "dry" conditions.

Calibration of the pneumotachograph is accomplished with the help of a piston of exactly 1 l gas capacity at room temperature which is pushed through the measuring head, while the resulting flow intensity (pressure difference in cm H_2O) is stored via the analog-digital converter in the same time interval. The calibration constant in ml/minute is

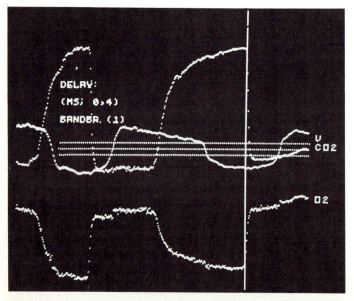

Fig 192   Recording of two respirations. Curves marked O_2 and CO_2 reproduce course of corresponding gas partial pressures. V denotes the pneumotachogram. Respiration is expected only when pneumotachogram surpasses required threshold which has been set with the knob (1) of computer. Delay (see text) is zero. It can be seen that end of expiration (vertical line) does not agree with passage of pneumotachogram through the zero point.

then the value with which the sum of all the stored values must be multiplied to obtain the value of 1.0.

Using programmed questioning via the display, the name, age, weight and other information on the subject, as well as the barometric pressure and room temperature, will be obtained and stored with the data, so that the values on the magnetic disc will be easy to find later on.

8.3.4 Determination of the Delay Periods of the Apparatus

Figure 192 should clarify the additional problems which occur. In this figure, information from several respiratory cycles has been stored in the computer and made visible on the display. Three curves can be seen: the curve labeled V shows the intensity of respiratory flow and the curves labeled O_2 and CO_2 reflect the pattern of the corresponding gas concentrations. From equation 84, it is clear that time-synchronous sections of the pneumotachogram and the gas concentrations must always be multiplied with one another and then summed together. Although the pneumotachograph is technically a pressure differential

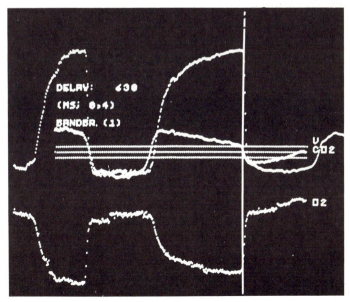

Fig 193   Synchronization of gas pressure curves and pneumotachogram. Passage through zero point is now manually (knob 4 and zero) synchronized with end of respiration. Value for delay now appears on picture screen.

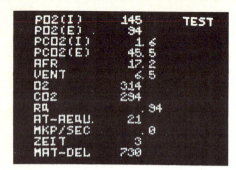

```
PO2(I)      145      TEST
PO2(E)       94
PCO2(I)       1.6
PCO2(E)      45.5
AFR          17.2
VENT          6.5
O2          314
CO2         294
RQ            .94
AT-AEQU.     21
MKP/SEC       .0
ZEIT          3
MAT-DEL     730
```

Fig 194
Information on display which indicates measured values in their correct dimension after each respiration (p in mm Hg., Afr in breaths per min, vent. in l/min, $\dot{V}O_2$ and VCO_2 in ml/min, time in sec, MAT-delay in msec).

sensor which operates with practically no delay, the transport of gas from the sampling site to the analyzer of the mass spectrometer requires between 500 msec and 1 sec, during which the form of the curve is not changed. To find the lag time between the two sets of apparatus (delay), it is best to proceed such that at the beginning of inspiration, the concentration indicated must drop rapidly to inspiratory values. At the same time, the respiratory flow must have a value of 0. To equate for the delay, values for respiratory flow intensity are stored at a fixed but adjustable lag time and then are called back for further use. The delay value can be determined by the vertical line graphically drawn in, in which the drop in concentration is synchronized with the zero flow of the flow intensity curve, as shown in Figure 193. The delay in msec is simultaneously displayed in the upper left of the screen. The delay is not constant if, for example, during heavy physical exercise expired air becomes warm, because then the viscosity relationships in the Teflon tube and in the mass spectrometer capillaries are altered. The program developed by us automatically corrects the delay time during the investigation. For control purposes, during the recording each delay time is shown on the display to the person in charge of the experiment (Fig 194).

8.3.5 Determination of the Duration of Breathing

Another technical recording difficulty is the precise determination of zero flow of the respiratory flow-intensity curve. Highly sensitive manometers have a tendency to "drift", i. e., their zero point shifts over a prolonged period. Therefore, an automatic zero point correction is inserted into the program, whose principle is shown in Figure 195. Within the two tubes which lead from the measuring head of the pneumotachograph to the differential manometer, two magnetic valves are built in which are connected with the relay register of the computer. After the determination each minute, a short impulse is

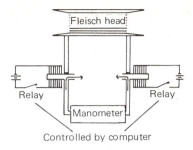

Fig 195 Automatic zero point correction of the pneumotachograph.
Within adjustable time intervals,
manometer is connected to exterior
air by switching on relays by the
computer. In this way, electric zero
point is recalibrated each time.

produced by the program via the corresponding relay; this impulse
connects the manometer with the external air. Because the differential
manometer therefore indicates the value of zero for about one second,
the difference between the indicated voltage and the true voltage of
zero is recorded by the computer; this difference will be added to the
actual values obtained during further data collection. The zero-point
drift is thereby eliminated.

Another difficulty is that the subject often does not breathe in an
uninterrupted manner at the range of zero flow, but breathes with
several accumulative oscillations. Under resting conditions, this can
have a disturbing effect because the total design of the program is
based on the exact determination of the inspiratory and expiratory
phases. Therefore, an adjustable response threshold is provided (Fig
193). Its size can be determined from the distance between the zero
line and the line parallel to the abscissa. Inspiratory or expiratory
phases will be evaluated only when the pneumotachogram has passed
through the range of these two lines.

8.3.6 Measurement and Calculation of the Spiroergometric Variables

The course of the measuring program occurs in the "interrupt mode";
this means that the basic program operates at several levels and that
these individual levels have varying priorities. In addition, the central
processor works parallel to a floating-point processor. In the present
case, the lowest priority is that of the display program, which continually displays the results of the breath-by-breath analysis on the screen
(Fig 194). This program will operate when the computer has nothing of
more importance to do. A digital computer works so rapidly, however,
that the interruption cannot be perceived with the eyes. The analog-
digital conversion initiated via the quartz clock, on the other hand, has
the highest priority. Between these priorities are the subprograms for

the automatic subsequent adjustment of the delays, for expressing the results and for storing the data on the magnetic disc for later off-line analysis.

The actual measuring process operates in the following manner: Each value for respiratory flow intensity is continually converted within the interval of 10 msec and stored. If the previously determined delay is 700 msec, then about 70 values which have been stored are used to summate their content on a register which had a value of zero at the beginning of the respiratory cycle. The same content of the 70 stored values are multiplied with the value of the simultaneously converted O_2 concentration minus the inspiratory concentration (after correction with the calibration factor) and summated onto another storage register. The same occurs with the CO_2 and N_2 values. This process is repeated at intervals of 10 msec, until such time as the delayed pneumotachograph curve once more shows a zero flow. At this time, it is determined whether this zero flow is at the beginning or at the end of the respiratory curve.

Thus, at each zero flow the content of the summation registers in question always shows either the inspired or expired respiratory volume or the amount of O_2, CO_2 or N_2 inspired or expired. The continuous digital integration spares the complicated calculation of the shrinkage in inspired air, as would be necessary in a purely expiratory measurement. The inspired volume, reduced to that of the expired volume, thus produces for each measurement either the consumption or the elimination, respectively. Before the values can be relayed to the display or printed in full, they must still be converted to BTPS or STPD and displayed in the proper dimension.

Such derived values as ventilatory equivalent are simultaneously calculated but not stored because they can be reproduced each time from the basic values. The advantage of integration over the total breath is that intake or elimination of gas volumes can be determined independent of the concentration of the corresponding gases in the external air.

8.3.7 Determination of the End-Tidal Gas Pressures and Adjustment of the Inspiratory Gas Pressures

The expiratory CO_2 pressure is measured in such a way that after continuous digitalization and storage of the mass spectrometer data over a single breath, the maximal value for CO_2 is sought. After multiplication with the appropriate calibration constants, it is given as the end-tidal pressure and stored. The minimal pressure represents the inspiratory CO_2 pressure and is also presented.

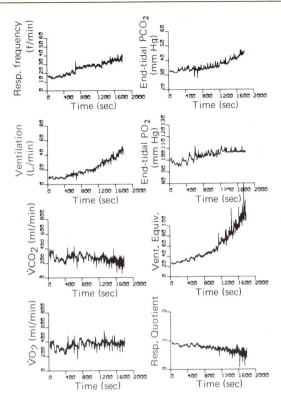

Fig 196   Illustration of resting experiment in which inspiratory CO_2 concentration (not shown) was continually increased from 0 to 6% within 30 minutes.

O_2 pressure is obviously determined in the same way, except that here the maximal value is the inspiratory pressure and the minimal value is the end-tidal pressure.

To test the reactions of the body to hyperoxia, hypoxia or hypercapnia at rest or during exercise, variable inspiratory concentrations of O_2, CO_2 or even other gases which can be determined with the mass spectrometer can be controlled. They can be preprogrammed to a constant value or even altered according to calculable functions. The regulatory principle is presented in Figure 196. An adjustable compressor delivers a variable amount of air, which is led past the inspiratory side of the pneumotachograph by means of a T-shaped tube. About one meter before the pneumotachograph head, the tube has another T valve. If hypercapnia is desired, for example, a defined amount of pure CO_2 gas is added. The amount is determined by a needle valve, which is adjusted by a program-controlled step motor. The optimal control conditions are obtained with the help of an integral regulator (Suppl, p 305). In this way, the adjustment motor is so regulated that its rotational speed is proportional to the present deviation between the actual and desired values. The desired value (set point) is determined by the program, whereas the actual value is the measured inspiratory gas concentration during each breath. For example, if the inspiratory set point is 4% CO_2 but the actual value is only 2%, then the motor would be rapidly adjusted; the more it approaches the required concentration, the slower will be the adjustment of gas flow. Excessive oscillations are avoided with this principle. Each gas concentration to be controlled requires a specific regulatory adaptation.

8.3.8 Programmable Performance Adjustment

For programming work loads, we use a modified Müller bicycle ergometer (Fig 42). In its original form, work load was programmed by the depth of a slotted magnet that is adjustable by a small hand crank. Because the original brake magnet was too weak to determine the maximal O_2 intake of a well-trained athlete, a stronger magnet was applied and the system was recalibrated. Instead of the cranking mechanism, a step motor was controlled by the computer; at 60 rpm, one step of the work load adjustment of 0.25 kpm/second corresponded to about 2.5 W. The maximally adjusted speed was about 400 W/second, so that rapid transitional and sinusoidal work load functions could be prescribed with the aid of the computer program. The actual work load value (Fig 193) is continually shown on the display. The extent of the transitional change and the duration of the work load are given in a data table. This arrangement also permits recording of the dynamic responses of the cardiopulmonary system.

8.3.9 Off-Line Representation of the Results

The investigation results can be given either in digital form (numbers) or in analog form (curves). The issuance of numerical values during the study can result in such a way that the above-mentioned values can be printed on a line printer breath-by-breath. Immediately after the study, the results can also be recorded on an incremental plotter as curves, with descriptions on the abscissa and the ordinates (i. e., ready for publication or storage). Figure 101 shows a study on a physical education student, during which the work load was progressively increased to the point of exhaustion. Each level had a duration of 3 minutes and a load of 7.5 kpm/second. The number of pedal revolutions was 75 per minute. Oxygen intake reached a maximal value of 4.5 l/minute. The other diagrams show the typical pattern of the associated variables. Figure 197 is a recording of the respiratory response curve to CO_2, where the inspiratory CO_2 concentration (not shown) was continually increased from 0 to 6% within 30 minutes. This figure shows that the method permits the adjustment of desired inspiratory gas pressures without disturbing the recording.

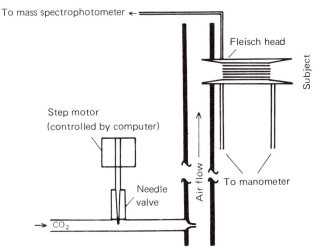

Fig 197   Adjustment of inspiratory gas pressures (shown for CO_2). Desired value is preprogrammed. Actual value is measured via inspired gas pressure. Step motor which controls needle valve is adjusted for speed and direction depending on difference between actual and desired value.

Terminology

Absorption: Uptake of gases in fluids or solid objects, synonymous with the formation of a true solution of gases. Also used as an expression for the uptake of visible radiation by matter, where the light and radiation energy is converted into chemical energy and heat energy.

Acclimatization: Adaptation to altitude or climatic conditions.

Acetylcholine: Hormone found in the tissue which primarily transmits the effects of the parasympathetic nerves.

Acidosis: Increase of hydrogen ion concentration in the blood.

Actin: Specific muscle protein.

Action current frequency: Number of action potentials (spikes) per unit of time which travel along a nerve fiber or a muscle fiber.

Actomyosin: Specific protein which is formed during muscle contraction from the muscle proteins actin and myosin.

Adaptation: Adjustment – in physiology, it is especially the temporal decrease in excitation of a receptor or a synapse during a constant stimulation.

Adenosine Triphosphate (ATP): Energy-rich phosphate which yields the primary energy for muscle contraction and drives the ion pumps.

Adequate stimulus: A stimulus which is specific to the corresponding receptor, as opposed to an inadequate stimulus. Even in the case of a punch in the eye, the inadequate stimulus transmits the sensory perception of light.

Adrenalin: Epinephrine.

Adrenosympathetic: Pertaining to effects which the adrenal medulla and the sympathetic system exert together.

Aerobic: Pertaining to metabolic processes which occur only in the presence of oxygen.

Afferent: Pertaining to stimulus transmission toward the central nervous system.

Agonist: The muscle which exerts an action reciprocal to that of the antagonist.

Alactacid: Without the formation of lactic acid.

Aldosterone: Hormone of the adrenal cortex which mainly affects electrolyte balance.

Alkalosis: Reduction of hydrogen ion concentration in the blood.

Alveolar concentration: Gas concentration in the area of the pulmonary vesicles.

Alveoli: Pulmonary vesicles.

Amino acid: Organic acid, in whose carbon skeleton one or more hydrogen atoms are replaced by the amino group NH_2.

Amino acid sequence: Characteristic serial composition of different amino acids. It is protein-specific.

Ampholytes: Chemical compounds which split off H^+ ions as well as OH^- ions, depending on their hydrogen ion concentration, i. e., they can dissociate as an acid or as a base.

Anaerobic: Pertaining to metabolic processes which take place without the participation of oxygen.

Anastomosis: Cross-connection between blood vessels (also between arteries and veins), between lymph vessels, or between nerves.

Androgens: Male sex hormones.

Anions: Negatively charged ions.

Anoxia: Absolute oxygen deficiency in the blood or in a tissue.

Antagonist: See *Agonist*. The reaction which initiates a reciprocal effect is called antagonistic.

Antidiuretic hormone (ADH): Hormone which is formed in the hypothalamus and liberated via the hypophysis and which affects the reabsorption of water in the kidneys.

Aortic and carotid bodies: Plexi near the aortic arch and carotid sinus, respectively, which are important for the respiratory regulation (oxygen deficiency and alterations in hydrogen ion concentration) of the blood.

Apathy: Psychic lack of interest.

Apneic diving: Diving without apparatus while holding the breath.

Arrhythmia: Irregular succession of heart beats.

Arterioles: Blood vessels in the area of the terminal blood stream which lie between the smallest arteries and the capillaries. They do not have complete vascular musculature.

Arteriovenous difference: Difference in measurements between the artery and the vein.

Aspiration: As a rule, suction of foreign material from the trachea.

Assimilation: Transformation into body substances of material taken from a living organism (anabolism, metabolic synthesis).

Atrium: Upper chamber of the heart.

Atrophy: Shrinkage of tissue.

Atropine: Poison of belladonna which paralyzes the parasympathetic nerves.

Autogenic training: Procedure for the voluntary modification of autonomic reactions.

Autonomic: Capable of spontaneous activity.

Autonomic system: Part of the nervous system which is generally responsible for involuntary reactions and regulations. It consists of the sympathetic and the parasympathetic nervous systems.

Autoregulation: Regulation of blood flow without nervous influence. It occurs due to the fact that smooth musculature changes its level of tension when it is stretched.

Auxotonic: Pertaining to the form of contraction, during which length and tension are both changed.

Axial flow: Central flow during laminar flow, as opposed to the flow at the edges.

Axon: Usually a long appendage of a nerve cell. Many axons form the peripheral nerve.

Bainbridge reflex: A reflex described by Bainbridge which causes a rise in heart rate during extreme stretching of the right atrium.

Baroreceptors: Stretch receptors in the area of the aorta and the carotid sinus which measure the stretching of the vascular wall and transmit information about that pressure.

Barotrauma: Injury due to pressure, e.g., during diving.

Bathmotropy: Alteration of the stimulus threshold of the myocardium.

Biomechanics: Science of the application of mechanical laws to the organism.

Biopsy: Observation and analysis of tissue samples taken from the living organism.

Bradycardia: Low heart rate.

Calorie, large (kcal): The amount of heat required to warm 1 kg water from 14.5 to 15.5 C.

Calorimetry: Measurement of the production or loss of heat.

Carboanhydratase: Enzyme which accelerates the hydration of carbonic acid.

Carotid sinus: Protrusion of the internal carotid artery near the separation of the common carotid artery into the internal and external branches. The stretch sensors which monitor blood pressure are present here.

Caudal: Belonging to the tail, toward the posterior end. A term derived from animal anatomy. Antonym: cranial, i. e., lying toward the head.

Cerebrospinal fluid: Fluid of the brain and spinal column.

Collateral vasoconstriction: Increased blood flow in the working regions and reduced blood flow in the resting regions of the body.

Colloidal-osmotic pressure: Also called oncotic pressure. That portion of the osmotic pressure which is produced by protein molecules in the blood.

Compressibility: The ability to be squeezed together; fluids are not compressible, gases follow the Boyle-Mariotte Law ($P \cdot V$ = constant).

Concentration: Amount of a molecule per unit of space in a defined phase.

Convection: Transmission, especially the transmission of energy transported from the smallest portion of a flow.

Cupula: Sensory organ in the semicircular canals whose adequate stimulus is rotational acceleration.

Cybernetics: In biology, the study of biologic servocontrol loops and of information transmission within the human body.

Cytochromes: Important iron-containing hemin proteins. As catalyzers of cell respiration in all tissues, they make possible the utilization of oxygen via electron transport.

Decompression: Sudden drop in pressure in the environment of the organism.

Delirium: Mental disturbance; those who suffer from this do not recognize their environment, are disoriented and are under the influence of sensory illusion, which induces them to perform incomprehensible acts.

Depolarization: The breakdown of membrane potential as a result of stimulation.

Diastole: Portion of the cardiac cycle, during which the heart is not contracting.

Disaccharide: Double sugar, e. g., beet sugar.

Dissimilation: Totality of the metabolic processes through which simpler compounds or end products of metabolism are formed from higher organic compounds (antonym: assimilation).

Distal: Toward the periphery.

Dromotropy: Influence on the transmission time of the heart.

Effective temperature: Compound climatic index formed from temperature, wind velocity and humidity of the air.

Effector: Activating organ, e. g., muscle, gland.

Efferent: Pertaining to transmission of a stimulation from the central nervous system toward the periphery.

Elasticity: Ability to store deformation energy in a reversible way.

Electromechanical coupling: Biologic process which produces contraction of muscle fiber on stimulation.

Embolism: Carrying of a blood clot, foreign object, amniotic fluid, fat droplets or air through the blood stream, with the end result of vessel obstruction.

Empirical: Based on experience; in scientific usage, those results which are based on observation, measurement or experiment.

Emulsion: Product of two nonmixable fluids, where one fluid represents the disperse phase in the form of tiny droplets of 1–50 μm diameter; these are suspended in another fluid (closed phase).

Endogenous: Originating from within.

Endolymph: Fluid in the membranous labyrinth of the ear.

Endothelium: Thin, single-layered lining of flat cells on the inner side of blood and lymph vessels.

Enzymes: Proteins produced in living cells which accelerate slow-moving chemical reactions for degradation and synthesis and also permit these chemical reactions to move toward the left.

Epinephrine: Hormone of the adrenal medulla; when released into the blood, it produces reactions similar to sympathetic stimulation.

Equilibrate: Put into balance.

Equivalent: Having the same value; physical: the mechanical heat equivalent means that 427 kpm corresponds to 1 kcal; chemical: molecular weight or atomic weight (in grams) divided by valence = mol/valence = gram equivalent = Eq.

Ergometer: Apparatus for measurement and adjustment of work or power.

Ergonomics: Science of industrial work.

Erythrocytes: Red blood cells.

Erythropoiesis: Formation of red blood cells.

Essential amino acids: Amino acids necessary for protein synthesis but which are not formed in the body from other amino acids.

Essential fatty acids: Fatty acids which are not capable of being produced in the body and must therefore be taken in from external sources.

Expiration: Breathing out.

Extracellular fluid: Fluid outside of the cells.

Glomerular filtrate: Fluid which is filtered out of the glomeruli of the kidneys. In the course of urine formation, most of the substances filtered out in water are reabsorbed.

Glycogen: Animal starch.

Glycolysis: Breakdown of glycogen or glucose without oxygen.

Hematocrit: Fraction of blood cells in total blood.

Hemodynamics: Study of the laws of movement of the blood, especially the dynamics of the heart and the vascular functions.

Heparin: Coagulation-inhibiting substance.

Homeothermic: Of constant temperature; this is the case with the core temperature of so-called warm-blooded animals (antonym: poikilothermic = cold-blooded).

Humoral: Having influence via the body fluids.

Hybrid: Interconnection of analog and digital elements in a computing element.

Hydrolysis: Splitting of chemical compounds by the addition of water.

Hydrostatic pressure: Pressure exerted by a column of water.

Hypercapnia: Increase in alveolar and arterial CO_2 pressure.

Hyperplasia: Increase in muscle diameter by the formation of new muscle fibers.

Hyperpolarization: Increase in membrane potential.

Hypertension: High blood pressure.

Hypertrophy: Increased strength of the muscle due to an increased diameter of individual muscle fibers.

Hyperventilation: Ventilatory factor which allows the CO_2 pressure in arterial blood to drop.

Hypocapnia: Reduction in alveolar or arterial CO_2 pressure.

Hypophysis: Pituitary gland.

Hypothermia: Undercooling.

Hypoventilation: Ventilatory factor which allows the CO_2 pressure in arterial blood to rise.

Indifferent temperature: Temperature at which the body requires no additional regulatory mechanisms to compensate for the effects of cold or heat.

Inotropy: Effect on the contractile force of the myocardium.

Insensible perspiration: Moisture which evaporates due to the leakage of water through the skin.

Inspiration: Breathing in.

Insufflation: Blowing of gaseous, fluid or powdery substances into body cavities and blood vessels.

Intermediary metabolism: Metabolism occurring in the median stage, i. e., neither the initial nor the final stage.

Interstitial space (interstitium): Space between cells.

Intrafusal: Within a muscle spindle.

Intravascular: Within the blood vessels.

Isobar: Line of equal pressure.

Isoelectric point: H^+ ion concentration at which acid and base groups of a molecule are equally dissociated.

Isohydre: Line of equal hydrogen ion concentration.

Isometric: Pertaining to a contraction with a constant length.

Isotherm: Line of equal temperature.

Isotonic: Pertaining to a contraction with constant tension development.

Isotropic: Having equal optical refraction characteristics.

Kinetosis: Movement disturbance, e. g., sea sickness.

Lactacid: With the formation of lactic acid.

Laminar flow: Form of flow in which the flow filaments run parallel to the direction of flow and where the axial or central flow runs faster than the flow at the boundary. In general, laminar flow has a parabolic velocity profile and obeys the Poiseuille Law.

Latent period: Period between the stimulus and the result of the stimulus.

Law of mass action: Basic law, according to which a chemical reaction in the gas phase or in solution never completely runs its course, but stops as soon as chemical equilibrium is reached.

Limbic system: In man, a system like a margin (limbus) of the brain stem and the surrounding cortical and subcortical regions, with fibrous connections to the nuclei of the diencephalon and midbrain. With each emotional stimulation (such as passion and emotion), the limbic system arouses the associated autonomic regulation; it is also concerned with actions for the preparation of motivational satisfaction.

Lipase: Fat-splitting enzyme.

Medulla oblongata: Prolonged medulla, the site of such important autonomic centers as the respiratory and circulatory centers.

Migraine: Sudden and periodic headaches, often appearing on one side.

Monosaccharide: Simple sugar, e. g., glucose.

Morphology: Science of the construction and organization of organisms and their components. It includes primarily anatomy, histology and cytology.

Movement stereotype: Automatic movement pattern.

Myoglobin: Muscle pigment which can transport oxygen similar to the manner of hemoglobin in the blood.

Necrosis: Cellular death.

Neuron: Nerve cell with its appendages.

Nomogram: A graphic representation of functions using curves within a plane. These can then be used to graphically solve mathematical problems whose treatment would otherwise require a great deal of effort.

Norepinephrine (noradrenalin): Sympathomimetic transmitting substance which is primarily formed at the sympathetic nerve endings.

Normoxic: Under normal conditions of oxygen pressure.

Oncotic: Colloidal-osmotic.

Orthostatic tolerance: Primarily the tolerance of the circulatory system to regulate against changes in body position.

Oxidation: Combination with O_2, withdrawal of H_2, withdrawal of electrons.

Parasympathetic: Pertaining to a part of the autonomic nervous system, also called the trophotropic system, which is primarily activated in the recovery phase.

Parenteral: By-passing the intestinal tract.

Partial pressure: In relation to the total pressure, ideal gases have the partial pressure which corresponds to their concentration in the gas mixture.

PD sensor: Proportional-differential sensor; in biology, receptors which are simultaneously affected by stimuli and changes in stimuli.

Pharmacology: Study of the effects of drugs.

pH value: Negative logarithm of the hydrogen ion concentration (example: $[H^+] = 10^{-7}$ mol/l corresponds to pH = 7).

Plastic: Pertaining to material which can change its form without storing the deformation work.

Pleura: Thoracic and lung membrane.

Pneumotachograph: Measuring apparatus for the determination of the intensity of respiratory flow.

Polysaccharide: Compound sugar, e. g., starch.

Polysynaptic: Traversing many synapses.

Proprioceptors: Receptors by which inherent organ and system reflexes are initiated.

Proximal: Toward the center.

Pulse pressure: Difference between the systolic and diastolic blood pressures.

Receptor: Biologic structure for the conversion of a stimulus into a biologic excitation.

Refractory period: The period following a stimulation, during which the membrane is not or only slightly excitable.

Regression: Statistical relationship between an independent and a dependent variable.

Renshaw cells: Inhibitory intermediary neurons composed of monosynaptic and polysynaptic reflexes.

Respiratory chain: Enzyme system in the tissues which permits the reaction of oxygen with the hydrogen formed.

Reticular formation: Plexus of nerve cells between the medulla oblongata and the thalamus.

Retina: Perceptive structure of the eye.

Selective permeability: Under this condition, the membrane represents a great obstruction for one type of ion and a small obstruction for another type.

Sensible perspiration: Evaporation due to sweat loss as a function of the sweat glands.

Specific heat: The specific heat of water is 1 (see kcal). Specific heat of other substances is given as a number proportional to that of water.

Spinal nerves: The spinal cord contains 31 pairs of spinal nerves. Each individual nerve results from the combination of 5–10 individual nerve fiber bundles of the existing anterior and posterior roots. Before combining to form the stem of the spinal nerve, the posterior root in the intervertebral canal has an oval swelling, the spinal ganglion. Each spinal nerve formed in this way contains sensitive, motor and autonomic fibers.

Splanchnic region: (From splanchnos = viscera, the region supplied by the splanchnic nerve). This is primarily an autonomic fiber which supplies glands and visceral muscles, as well as the sensitive fibers coming from this region.

Standard bicarbonate: Bicarbonate concentration in blood plasma which is equilibrated with a mixture of oxygen and 40 mm Hg P_{CO_2} at 37 C.

Static exercise: That physiologic work which is performed when a weight is held. In this case, the external physical work = 0.

Steady state: A balance between inflow and outflow or between assimilation and dissimilation.

Stenosis: Narrowing.

Stoichiometry: Study of the quantitative composition of chemical compounds and their relative amounts during chemical reactions. In stoichiometry, the masses or volumes (in the case of gases and solutions) of reacting substances are placed in relationship to one another.

Stress: Collective term for overloading an organism through external and/or internal stimuli which surpass normal levels, as well as the totality of all the inherent physical and specific adaptational and defense reactions that appear.

Subcortical: Beneath the cerebral cortex. Subcortical reactions are understood to be reactions which occur without participation of the cerebral cortex.

Switching on of the disturbance variable: Technical regulatory term. In order to span the slow reactions of a regulator, the actual value is briefly uncontrolled and moved in the direction of the set point. After the effect of switching on the disturbance variable fades away, the regulator once more assumes control.

Sympathetic: Pertaining to a part of the autonomic nervous system, also called the ergotropic (performance-increasing) system because it is primarily activated during performance, stress, etc.

Synapses: Junctions between two excitable cells, e. g., two neurons, but also between nerve cells and muscle cells.

Systole: Period during which the myocardium is contracted.

Tachycardia: High heart rate.

Teleologic: In biology, a philosophical term which strives to explain natural processes according to their sense and purpose.

Terminal: Situated at the end.

Testosterone: Male sex hormone belonging to the group of androgens.

Tetanus: In physiologic terms, the normal compound spasm form of muscle contraction. It is produced by the summation of excessive individual twitches (superimposition), during which the second muscle twitch is placed on that of the first when the temporal distance between the two initiating stimuli is less than the duration of the twitch.

Threshold stimulus: Magnitude of a stimulus which is just capable of initiating an excitation.

Tonus: Tension.

Torr: Unit of pressure, identical with mm Hg (mercury) column.

Trachea: Windpipe.

Transmural pressure: Difference between internal and external pressure of a blood vessel.

Turbulent flow: Flow with the formation of eddies and whirls; antonym: laminar flow.

Vagovasal syncope: Characteristic collapse of the circulatory system when orthostatic regulation breaks down.

Vasoconstriction: Narrowing of blood vessels.

Vasodilatation: Widening of blood vessels.

Vasomotor nerves: Vascular nerves which supply the smallest arteries.

Ventilatory equivalent: Ventilation (respiratory minute volume) divided by oxygen consumption ($\dot{V}_E/\dot{V}_{O_2}$).

Ventricle: Chamber or cavity.

Viscosimeter: Apparatus for measuring the viscosity of a fluid.

Index